DIGITAL BEHAVIOR

Evidence-Based Digital Design

UMBERTO LEÓN DOMÍNGUEZ

University of Monterrey

CAMBRIDGE
UNIVERSITY PRESS

CAMBRIDGE
UNIVERSITY PRESS

Shaftesbury Road, Cambridge CB2 8EA, United Kingdom

One Liberty Plaza, 20th Floor, New York, NY 10006, USA

477 Williamstown Road, Port Melbourne, VIC 3207, Australia

314–321, 3rd Floor, Plot 3, Splendor Forum, Jasola District Centre, New Delhi – 110025, India

103 Penang Road, #05–06/07, Visioncrest Commercial, Singapore 238467

Cambridge University Press is part of Cambridge University Press & Assessment, a department of the University of Cambridge.

We share the University's mission to contribute to society through the pursuit of education, learning and research at the highest international levels of excellence.

www.cambridge.org
Information on this title: www.cambridge.org/9781009568333

DOI: 10.1017/9781009568296

First published 2025

A catalogue record for this publication is available from the British Library.

Library of Congress Cataloging-in-Publication Data
NAMES: Domínguez, Umberto León, author.
TITLE: Digital behavior : evidence-based digital design / Umberto Leon Dominguez.
DESCRIPTION: Cambridge ; New York, NY : Cambridge University Press, 2025. | Includes bibliographical references and index.
IDENTIFIERS: LCCN 2024022047 | ISBN 9781009568333 (hardback) | ISBN 9781009568326 (paperback) | ISBN 9781009568296 (ebook)
SUBJECTS: LCSH: Human-computer interaction – Psychological aspects. | User interfaces (Computer systems) – Design. | User-centered system design. | Internet users – Psychology. | Information behavior.
CLASSIFICATION: LCC QA76.9.H85 D64 2025 | DDC 004.01/9–dc23/eng/20240923
LC record available at https://lccn.loc.gov/2024022047

ISBN 978-1-009-56833-3 Hardback
ISBN 978-1-009-56832-6 Paperback

DIGITAL BEHAVIOR

In a technologically advanced and competitive landscape dominated by major tech companies and burgeoning start-ups, the key asset lies in boosting monthly active users. Traditionally, product design has relied on fragmented insights from personal experience, common sense, or isolated experiments. This work endeavors to establish a theoretical framework for predicting and influencing the digital behavior of technology users. Drawing on over a century of scientific research in behavior, cognition, and physiology, this presents a comprehensive approach to customizing digital stimuli. The objective is to enhance user interactions with digital and virtual environments. Through real and cost-effective examples, diagrams, and formulas, the text offers theoretical knowledge and a practical methodology to elevate digital product designs, setting them apart from the competition. With the potential to reshape the digital design landscape, this book emerges as a game-changer, promising to revolutionize how digital products and services are conceived and delivered.

DR. UMBERTO LEÓN DOMÍNGUEZ is Professor of Psychology at the University of Monterrey, contributing significantly as a member of the Artificial Intelligence Research Group. In addition to his academic role, he is the founder and director of the Laboratory for Human Cognition and Brain Studies at the university. Dr. León Domínguez also holds the position of Director of Artificial Intelligence at FREIGHT Technologies Inc., a prominent transnational logistics technology company listed on the NASDAQ.

"We can only command nature by obeying her."

Bacon (1620)

Contents

Figures

Tables

xi

Preface

Transhumanism posits technology as an evolutionary mechanism capable of accelerating human development. The current technological boom, marked by the emergence of novel artificial intelligence (AI) algorithms and techniques such as diffusion or transform models, represents a potential inflection point in this process. The increasing pervasiveness of technology is transforming human behavior and reshaping our social landscape. Traditional face-to-face interactions are being replaced by digital alternatives, as seen in the shift from physical shopping to online delivery, from outdoor play to virtual gaming, and from in-person dating to app-based connections. This transition necessitates the design of digital spaces that align with both the physical principles of cause-and-effect and the cognitive and behavioral principles of human psychology. Despite the recent emergence of dedicated computing degrees, the design of digital products and services often lacks a robust epistemological framework. Today's designs are based on a limited understanding of digital human behavior, beyond a handful of disconnected theoretical rules applied by designers, programmers, and some cognitive psychologists out of inertia, but without a theoretical framework explaining how and why we behave in the digital world. Fortunately, individual digital behavior is governed by the same principles as natural behavior, which can be explained by the same behavioral and cognitive principles. We do not need to reinvent the wheel. The expectations and consequences of our behavior shape our next interactions with digital products and services, making the understanding of these regulating principles imperative for professionals involved in designing and developing these technologies.

This book pursues three fundamental goals. The first is to create an epistemological framework that can serve as a reference for the design of digital products and services. Psychology has studied human behavior and cognition for over 100 years, and, fortunately, behavior in digital environments follows the same principles as behavior in natural (nondigital) spaces. This work compiles various scientific evidence explaining human behavior and cognition, relates it to real examples of different digital products, and finally, proposes a reproducible methodology for their design and development. Perhaps the most complex aspect of the book has been attempting to unify the behaviorist and cognitivist visions, two theoretically opposed psychological models, yet converging in practical reality. The second goal, a consequence of the first, is to professionalize and open the digital design field to psychology. This professionalization and opening aim to fill a gap in the empirical study of digital product and service design. This book proposes a framework that can apply a hypothetical-deductive methodology to propose hypotheses and test them, advancing the study of digital behavior design. Aligning the various physiological and psychological states of users with the administration of different digital and virtual stimuli in these technological spaces is crucial today, especially given the current boom in artificial intelligence, virtual, and augmented reality. This new specialty within psychology could be considered a sub-branch of human factor and engineering psychology, primarily tasked with finding the functional relationships between user needs, digital tool design, and need satisfaction through the consumption of reinforcers. As I tell my students, digital behavior designers are reinforcement designers. To design them properly, we must understand the dynamics between users' physiological, behavioral, and cognitive states. The last goal of this book is related to the ethics of designing and developing digital products and services. Often, designs not based on scientific evidence can lead to the abuse of certain elements that foster compulsive behaviors in vulnerable user populations. Satisfying certain needs through these tools can cause users to perceive them as sources of satisfaction, deregulating behavior control. This is a current problem for many families, to the extent that the first lawsuits against technology companies for designing products to "hook" users are beginning to emerge. This book also warns of these catastrophic consequences and proposes certain ethical and responsible behaviors that a digital behavior designer could perform.

Finally, this book is also the culmination of a personal project. As a doctor in neuropsychology, I have nearly twenty years of studying the brain and cognition, in addition to innovating and developing various technological and digital products, in which I had intuitively applied knowledge from neuropsychology. During this time, I realized that this knowledge from the neurosciences, behavioral sciences, and cognitive sciences was completely overlooked by those responsible for designing and developing digital products and services. However, when explained to them, they were amazed by the explanatory power of scientific explanations and the creativity it brought to digital design. Therefore, I decided to create a university class on Digital Behavior to teach my students how psychology is useful for digital and virtual design, and the reception was incredible. Many approached me asking where they could continue their education. Unable to find any text explaining what was discussed in class, I decided to write this book and make it available to the academic and professional community. However, I am aware that this project does not end here but marks the closure of one stage and the beginning of another. This new stage aims to consolidate, through experimental evidence, the design of digital behaviors. We may be at a point of no return for humanity, where technology and the human can merge into a new societal model where physical, technological, and digital realities intertwine to shape our new civilization model. Therefore, the professionalization of ergonomic design in technology and digital spaces is needed, so this techno-evolutionary transition emerges from a solid, scientific, and, above all, human base.

Abbreviations

CR	conditioned response
CR$^{\varnothing}$	no emotional conditioned response
CR^{AP}	appetitive conditioned response
CR^{AV}	aversive conditioned response
CS	conditioned stimulus
CS(+)	excitatory conditioned stimulus
CS(-)	inhibitory conditioned stimulus
ctxt	context
DBA	digital behavior analysis
DOD	drives and operants design
I-I-C Model	internal, interaction, and contingencies model
noS^{C+}	negative punishment
noS^{C-}	negative reinforcement
noS^{C-dd}	delayed direct negative reinforcer
noS^{C-di}	delayed indirect negative reinforcer
noS^{C-id}	immediate direct negative reinforcer
noS^{C-ii}	immediate indirect negative reinforcer
O^{D}	organism drive
NS	neutral stimulus
P^{B}	prior belief
R	operant response
S	stimulus or stimuli
S$^{\varnothing}$	absence of the expected stimulus
S$^{\Delta}$	delta stimulus
S^{C}	contingence stimulus
S^{AP}	appetitive stimulus
S^{AV}	aversive stimulus
S^{C+}	positive reinforcement
S^{C+dd}	delayed direct positive reinforcer

S^{C+di}	delayed indirect positive reinforcer
S^{C+id}	immediate direct positive reinforcer
S^{C+ii}	immediate indirect positive reinforcer
S^{C-}	positive punishment
S^D	discriminative stimulus
S^{NS}	neutral stimulus
US	unconditioned stimulus
UR	unconditioned response

CHAPTER I

Introduction to Digital Behavior

1.1 Interaction between Technology and Human Cognitive Skills

It is often said that humans have created technology, but seldom is it acknowledged that technology has also created the modern human. Technology has been used to change nature, but what we failed to notice is that as we were using technology, it was also changing us. Literally, the use of technological tools has altered our neural connections (Chun *et al.* 2018), and therefore, our way of seeing and understanding the world (Barr *et al.* 2015). With a certain evolutionary perspective and awareness, we continue to perform the same behaviors as our early homo sapiens ancestors, although now mediated by technology. Homo sapiens went hunting while we order Uber Eats; they went out into the village to find a partner while we search for one on Tinder; they spoke with their friends in person while we communicate through WhatsApp. They even painted in caves, and we continue to paint using Photoshop or new artificial intelligence (AI) algorithms like Midjourney, DALL-E, or Stable Diffusion. In other words, we continue to behave in ways that aim to accomplish the same tasks, only now within a much more complex and technology-mediated cultural environment. But what is technology? Technology can be defined in many ways, but as Albert Einstein said, "everything should be made as simple as possible, but not simpler." Therefore, to define what technology is, we can take a Vygotskian approach by considering it as cultural tools with the power to change the human mind (Vygotsky 1962, 1978). An extension of this definition can be taken from the philosopher Mario Bunge, who conceives of technology as objects or processes of possible practical value for individuals or groups, which have been constructed with the help of knowledge acquired through basic and applied sciences. Additionally, Bunge emphasizes a very important aspect that we do not always associate with technology – that it need not only be physical or chemical, but can also be biological (e.g., language) or social

(e.g., education or economics) (Bunge 1983). Ortega y Gasset also contrib-
utes to the debate of defining the concept of technology by describing it as
a human reaction toward nature or their personal circumstances, in a way
that leads to the design and construction of a tool or device that mediates
between them and nature (Ortega y Gasset n.d.). Therefore, technology
could present the following attributes:

– Functional: Technology is an object, product, device, cultural artifact,
  or tool (all synonyms) oriented to solve a need, whose material nature
  may be physical, biological, or cultural.
– Reactive: Technology arises from the biopsychosocial impulse to solve
  an individual, group, or community need. If it succeeds in solving the
  need, this technology acquires value.
– Scientific: Technology is a cultural product that is born and
  updated through the transmission and accumulation of cultural
  knowledge.

If we review the history of our species, we can observe how tools
have influenced the transformation of our species and culture, from
the invention of fire to the creation of artificial intelligence algorithms that virtually think for us. It is possible that in the last 100
years more technological progress has occurred than in the rest of
human history. This technological advancement is a violent process,
so much so that it changes our environment and literally changes us
through changing our brains. The brain is a large organ whose
functions require a high metabolic expenditure for its operation
(Aiello & Wheeler 1995). Once an adequate diet was achieved for
the maintenance of its energy costs (Gupta 2016), the brain managed
to develop certain higher cognitive skills through overcoming ecological (Clutton-Brock & Harvey 1980; Rosati 2017), social (Dunbar
1998), or cultural (Moll & Tomasello 2007; van Schaik & Burkart
2011) challenges. Although it seems that social challenges were the
ones that caused or triggered a cognitive explosion, recent research
carried out by Mauricio González Forero and Andy Gardner of the
University of St Andrews proposes that it was ecological challenges
(human vs. nature) that led to an increase in brain size, and
therefore, the development of cognitive skills (González-Forero &
Gardner 2018). In this regard, one theory that can explain the
particular development of human cognitive skills as overcoming
ecological challenges is proposed by Nobel laureate Herbert Simon
of Carnegie Mellon University, known as "bounded rationality"

(Simon 1957). Herbert suggests that cognitive development is neither greater nor lesser, but depends on one's own cognitive limitations and environmental challenges. For example, animals that have easy access to food and/or lack predators do not need to develop complex strategies for obtaining their food, while animals with restricted access to food and/or with threats of being hunted by another predator need to develop skills that allow them to generate strategies for survival. In humans, community organization and the need to obtain resources for the group led to facing different challenges with the cognitive skills they had. The development of technological tools led to the release of new sources of supply and social organization, and therefore, also posed new challenges that promoted the phylogenetic development of executive skills through the reorganization of the prefrontal cortex of the brain (Barrett *et al.* 2020).

To explain the processes by which technology is able to change our neural connections, we will start with Vygotsky and his sociocultural theory of cognitive development (Vygotsky 1978a). Vygotsky suggests in his theory that the cognitive development that occurs from childhood directly depends on our interactions with the elements of the environment that surround us. Specifically, he points out the importance for our development of symbolic cultural tools, such as language or, in a more contemporary case, the modern digital interfaces of smartphones. These person–tool interactions change our brain wiring, creating "specialized neural niches" in processing the information we obtain through interactions (Pinker 2010). The "neuronal recycling hypothesis" by Stanislas Dehaene offers a modern approach to understand the neuronal process that occurs in us when we acquire a new skill through cultural tools (Dehaene 2014b; Dehaene & Cohen 2007). This hypothesis indicates that interaction transforms or directly recycles the structure and connectivity of the neural groups responsible for processing the sensory-motor or symbolic information necessary to master the new technology, so that once these brain regions are transformed, we are able to process and manipulate more complex information (Amalric & Dehaene 2016; Dehaene 2009). Intuitively, we think that the greater the mastery of a skill, the more regions we activate, and this is false. A child who starts to perform simple arithmetic calculations such as addition and subtraction activates the same regions as an expert mathematician performing elaborate arithmetic calculations, only that the brain regions of the mathematician are much more connected (transformed) as a product of his experience in calculation (Cantlon *et al.*

2006; Cantlon & Li 2013; Izard *et al.* 2008). That is, it is the transmutation of a set of neural circuits that allows us to master the use of a new cultural artifact (Dehaene 2005, 2014a; Dehaene & Cohen 2007). A simple but curious experiment tries to explain how tools change the brain (Miller *et al.* 2018, 2019). In this experiment, an individual was placed in front of a curtain, unable to visualize the object that was behind it. The experimenter asked the subject to grab a stick and hit the object behind the curtain (without being able to see either the object or the stick), and then indicate with which part of the stick he had hit the object. All individuals were able to locate the area of contact of the stick with the object. The authors proposed that this occurred because the stick was processed by the brain as if it were an extension of the hand (extended cognition), as occurs in other animals, such as cats' whiskers when they brush against a wall. In some way, the brain fails to differentiate between the boundaries of the body and tools when they are in use, thereby integrating sensory information obtained through tools as if it were directly acquired by a sensory organ of the body. Consequently, this sensory information that arrives through tools seeks to adapt and find its own cognitive niche within the brain (Pinker 2010). In the specific case of the experiment just mentioned, the ability to sense tools as part of ourselves is due to the functional coupling between the biomechanical information from skin sensory receptors and the neural processing of higher brain regions (Miller *et al.* 2018). This mechanism of accommodating sensory information, as if acquired through our own organs, has enabled us to rapidly acquire the skill to manipulate new tools, a skill that in turn serves as a foundation for enhancing these tools (Martel *et al.* 2016; Miller *et al.* 2014).

All of this evidence is splendidly summarized in the cultural brain hypothesis, which proposes that through cultural learning, our brains have been selecting and enhancing our cognitive skills for manipulating and storing information obtained through the use of technologies (Muthukrishna *et al.* 2016, 2018). Among paleoscientists, there is a consensus that most technological innovations of Homo Sapiens have been developed due to the impact culture had on them (Colagè & d'Errico 2020; Bender & Beller 2019; Sterelny 2020). The impact of culture on humans can be assessed through two phenomena: cultural exaptation and the ratchet effect. The term "cultural exaptation" refers to how the manipulation of technological tools to solve novel challenges has generated a feedback loop of progressive and cumulative cognitive improvements by incorporating new information into cognitive schemas for tool use. The process of integrating this new information with existing cognitive schemas has transformed these schemas into new cognitive models for the use of these technological tools, thus facilitating the development of more advanced tools. In this way, cognitions evolved so that new tools could be created to solve new problems (d'Errico *et al.* 2018; d'Errico & Colagè 2018; Schlaudt 2022). For example, stones that were used for cracking nuts came to be used for cutting meat and as a weapon for hunting and attacking enemies. The creation of these new tools results in what is known as a ratchet effect, indicating the irreversible accumulation of these changes, both cognitively and culturally (Henrich *et al.* 2010; Tennie *et al.* 2009). As another example, the ability to name and write numbers was the cognitive basis for performing more complex mathematical calculations. That is to say, our cognition employed these cultural tools (naming and writing) as a foundation to transform itself and be capable of acquiring and processing more elaborate information like addition, subtraction, multiplication, and division (Miller & Paredes 1996). Similarly, the use of devices for problem-solving is able to transform our brain due to its plastic capabilities (Hecht *et al.* 2023). The direct consequence of this ongoing adaptation of our cognitive framework to new technologies has caused us to handle and create increasingly complex tools. This creates a loop where the creation of new tools changes our cognitive environment enabling us to create even more complex tools, thereby changing our cognitive environment yet again, and so on indefinitely. The most recent loop has succeeded in creating digital and virtual technologies, which currently represent the pinnacle of technological development. Many theories aiming to explain the human–culture relationship advocate that most cognitive processes that have allowed for cumulative cultural evolution (the ratchet effect), such as social learning, imitation, teaching, social motivation, theory of mind, and particularly reading, are solely cultural products, denying the importance of genes (Heyes 2014). On the other hand, other positions such as the culture–gene coevolution theory (Chudek & Henrich 2011), the dual inheritance theory (Richerson & Boyd 1978), and the cultural niche construction theory (Laland *et al.* 2000) advocate more for an interaction between culture and genes in human evolution.[1] Regardless of the interaction between genes and

---

[1] In this regard, there are segments of the human genome with a higher known rate of evolution, referred to as "human accelerated regions" (HARs). These are notably present in the human genome and have been associated with human neurodevelopment. More specifically, HARs act as regulatory elements that drive human genes such as PPP1R17, which enhances the development of cortical germline zones in primates, particularly in the human cerebral cortex (Girskis *et al.* 2021).

culture, both mechanisms allow us to adapt to the environment. While genes are a slow and long-term adaptive mechanism, brain plasticity and cultural objects (such as technological tools) allow for rapid environmental adaptation (Waring & Wood 2021). Indeed, human evolution would be determined by long-term gene–culture coevolution, where culture acts as an evolutionary vector. More recent research seems to validate this theoretical approach, maintaining that although genes significantly influence culture, culture is selecting them (Waring & Wood 2021). In this regard, culture has been overtaking genes as the evolutionary force in humans, which means that genetic traits are increasingly taking a backseat. There is even evidence to suggest that this trend is accelerating with the emergence of technology, reaching the pinnacle of human evolution when people are capable of biologically integrating with technology through soft devices.[2] This point at which humans merge with machines is known as "technological singularity," which would imply an exponential acceleration of evolution due to the artificial acquisition of biological and cognitive improvements (Petta Gomes da Costa 2019), with which interaction would occur through bioelectronic interfaces (Schiavone *et al.* 2020). The integration of individuals with technology is a transhumanist vision of human evolution and has sparked much debate in philosophical and academic circles, such as the dangers of artificial intelligence (Müller 2020), the possible emergence of a collective consciousness (O'Lemmon 2020), or even whether the new human resulting from the fusion of man and machine could be considered a new species (Enriquez & Gullans 2015). In this sense, technology, and its digital and virtual derivatives, seem to be determining the course of human evolution, making fields such as human factors or engineering psychology necessary to address not only the impact of technology on individuals and society but also to develop methodologies capable of generating ergonomic interfaces for individuals (APA 2014).

What is currently happening with the incessant emergence of new technologies? As Moore's law predicted, technology is progressing at an increasingly rapid pace, meaning our brain does not have time to accommodate new tools in their respective neural spaces or niches. Technology is advancing at a pace much greater than the time our brain has to structurally and connectively integrate it. This decoupling between the use of new technology and its integration with neural circuits is known as the "evolutionary mismatch theory," which could have a real impact on our health,

---

[2] Soft devices are a type of bioelectronic circuits that are compatible with human organic tissues (Wang *et al.* 2023).

such as an increase in the number of chronic diseases (Corbett *et al.* 2018; Pani 2000), or chronic stress (Brenner *et al.* 2015). This may be due to the fact that the current technological environment is very different from that with which our central nervous system evolved over millennia. Some scientists question whether the structural organization and neural plasticity of the human brain are incapable of absorbing the evolution of the environments we have created (Pani 2000). Even the biological cause of this evolutionary mismatch is hypothesized to lie in the mesocorticolimbic dopaminergic system, which could be overwhelmed by the dopamine release resulting from the continuous acquisition of reinforcers coming from digital environments (Wise & Rompre 1989). In this regard, the use of technology could induce changes in the dopaminergic system that could alter natural codes in reinforcement-aversion processes, sensorimotor mechanisms, and salient motivational cues; which have remained stable during human phylogenetic development, and which are necessary to associate the consequence of a behavior with a defined emotional state of the organism. This alteration of the dopaminergic system would affect its projections to the prefrontal cortex, which could affect intellectual capacities and emotional regulation of individuals (Pani 2000). Therefore, the ways in which technologies currently gratify us are far from those with which our nervous system evolved. We no longer need to "waste" several hours hunting to get food, but we go down to any bar or supermarket and are instantly gratified. Gone are the days when one had to visit a friend's house to inquire if they wished to go out; now we simply ask via WhatsApp and receive a response within seconds. We live in a society of instant gratification, which leads to an overactivation of the dopaminergic system and emotional deregulation. Nevertheless, technology has become a fundamental part of our lives and how we interact with our environment, with mobile phones being its primary standard-bearer. Mobile phones, with endless gratifications at our fingertips, have become an extension of our brains, fulfilling their needs swiftly and with minimal effort.

There is no doubt about the importance of technology in our lives, but we actually know very little about why and how we interact with it. There exists a vast body of knowledge historically generated from behavioral and cognitive sciences whose principles are perfectly transferable to the digital domain. Although classical authors such as Vygotsky, Skinner, Pavlov, and Hull lived in eras devoid of such technologies, they managed to produce ample theoretical and practical knowledge about human and animal behavior that could specifically explain human behavior within a digital environment, that is, digital behavior.

1.2   What Is Digital Behavior?

Behavior is the interaction of an individual in a physical environment. Conversely, digital behavior is the interaction of an individual in a digital environment (digital space). This distinction is crucial to emphasize, as it forms the axis around which the understanding of how and why digital behavior occurs revolves. Whereas in the physical reality, nature and culture provide us with stimuli, in the digital environment or space, it is technology that furnishes us with stimuli, which in this case are artificial and digitally natured. In the digital space, the technology interface sensorially presents us with stimuli, which can elicit the same effects as stimuli provided by nature or culture (Wu *et al.* 2023). This is a central point, as even though the stimuli impacting us are of different natures (natural vs. digital), both are sensory and, therefore, their effects can be predicted using the body of knowledge and tools already created from behavioral and cognitive sciences. That is, digital behavior design aims to use this knowledge and tools to predict possible user behaviors in digital environments and to develop ergonomic services that facilitate habit implementation. Through a study of the needs or goals of users when using a digital service, the digital behavior designer can apply psychological knowledge to identify and analyze the functional elements guiding people's behavior, thereby proposing an ergonomically sound digital technology solution aligned with their motivations. Digital behavior design can be described as a new subfield of human factors and engineering psychology. Human factors and engineering psychology is a science that studies human–machine interactions mainly to facilitate their use through error reduction and safety enhancement. This science draws from psychological knowledge for the design of products, systems, and devices used daily. The novel contribution of digital behavior design, and why it is included as a new subdiscipline within human factors, is that for the first time a theoretical and reference framework has been created for designing technological and digital products based primarily on the basic principles of associative learning. In this regard, digital behavior design would be a tool of human factors that would automate behavioral processes with which users would utilize various digital services or products. One of the subdisciplines that shares several epistemological areas with digital behavior design is human–computer interaction (HCI). The main difference, albeit nuanced, is that the body of knowledge guiding the former stems more from behavioral and cognitive sciences, while the latter originates from computer sciences.[3] Furthermore, there is also a significant epistemological difference. Professionals in HCI structure their work considering technology as an end-product, whereas digital behavior designers view technology as a means to satisfy a need. That is, habit formation and need satisfaction constitute the main epistemological core of digital behavior design.

1.3 Epistemological Principles of Digital Behavior Design

The design of digital behavior aims to establish the psychological foundations for the use of a technological service by anticipating user behavior in the digital environment. In this regard, personality and environment are the two best predictors of human behavior and, therefore, also of digital behavior (Cortina *et al.* 2022; Ozer & Benet-Martínez 2006). Digital behavior design primarily focuses on the environment, as it exerts more influence than personality in guiding specific behaviors (Cortina *et al.* 2022). In situations where environmental pressure is "strong," behavior is relatively uniform, regardless of an individual's personality. Conversely, in situations where environmental pressure is "weak," behavior is not constrained by the situation and may be more influenced by personality traits. This concept of the dependency of environmental pressure on behavior is known as "situational strength interaction," which underpins many psychological models of human behavior (Cortina *et al.* 2022). In this regard, behavioral and cognitive sciences have amassed over a century of scientific knowledge on how the environment affects behavior. These behavioral principles are scattered across various areas related to the development of digital services but are not structured and organized into a single discipline. Moreover, there is an urgent need for this knowledge to be applied in digital areas, given that we spend much of our lives there, and what we currently are and how we feel could be explained by understanding how we behave in the digital world. According to Skinner, the father of operant conditioning and a key figure in behavioral sciences: "We need to go beyond mere observation to a study of functional relationships. We need to establish laws by virtue of which we may predict behavior, and we may do this only by finding variables of which behavior is a function" (Skinner 1938, 8).

---

[3] Human-computer interaction (HCI) is a multidisciplinary scientific field dedicated to the study of interactions between humans and machines, drawing upon computer science, cognitive science, and human factors engineering.

Therefore, to access these behavioral principles, which will also explain digital behavior, the scientific method must be used. Like all science, the study of digital behavior is also governed by general axioms:

- Realism: indicates that reality exists independently of the observer.
- Intelligibility: reality, in addition to existing, can be understood.
- Localism: two objects distant in space-time cannot interact instantaneously, so only two objects close in space-time can affect each other instantaneously. This axiom is required for the principle of causality, which states that every event has a cause and an effect.
- Unitarity: for an event that has not yet occurred, the sum of all probabilities of that event occurring must equal one.

It is important to pause here and clarify a crucial point: for many psychologists, determinism is an axiom of science, but this is not entirely accurate (Earman 1986). Determinism is the idea or belief that the world follows the laws of cause and effect, not an undeniable axiomatic principle of science. Mechanical determinism summarizes the ideas of Hume and Descartes on causality (Márquez-Blanc 2012). Henceforth, the term "causality" will be used to refer to the relationship between two events. To explain and predict the causes of digital behavior, the theoretical principles proposed by behaviorism and cognitivism will be utilized, particularly those represented by the 4Es (embodied, embedded, enacted, or extended) (Carney 2020):

- Behaviorism: a set of principles that describe and explain behavior as a product of system–environment interaction.
- Cognitivism 4Es: a set of principles that explain that behavior is the result of interactions between our cognitive system and the environment, but mediated by mental representations[4] (or models of the world) that formed as a result of previous interactions.

Within cognitivism, the neuroconstructivist approach can also be considered, which suggests that the development of our brain is not an isolated process but is shaped by interactions with other systems over time; thus,

---

[4] In the scientific community, there is some debate surrounding the use of terms such as "mind" or "mental," as they are considered to be broad and vague. However, for certain sectors within psychology and neuroscience, mental representations are understood to correspond with specific neural representations in the brain. Various studies have found that neural representations tend to exhibit a topographical correspondence with the brain structures that process stimuli and generate behavior (Piccinini 2022b). For the design of digital behavior, mental representation is defined as "an emergent isomorphic system stemming from the integrative activity of various neural circuits, which allows for the cognitive operability of information encoded in their own circuits and that are subjectively experienced with the capacity to influence behavior."

behavior and cognition are the result of past interactions (Astle *et al.* 2023). Two other theoretical proposals that could have a significant impact on the study of digital behavior are Alekséi Leontiev's activity theory (Leont'ev 1978a) and Karl Friston's free energy principle (Friston 2010). Both proposals could be considered cognitivist. Activity theory aims to explain how tools (such as a hammer or a smartphone) regulate an individual's activity, which in turn transforms the individual's psyche (Leont'ev 1978b). This activity is modulated on one hand by the needs, motives, and tasks of the activity itself; and on the other hand, by the actions and operations necessary to perform the activity (execution), which have been defined based on previous needs, motives, and tasks (see Figure 1.1).

On the other hand, the design of digital behaviors is strongly influenced by the "free energy principle," specifically its application to neuroscience known as the "theory of predictive coding," which can be situated within the computational cognitive models of the human mind (Friston 2010). This

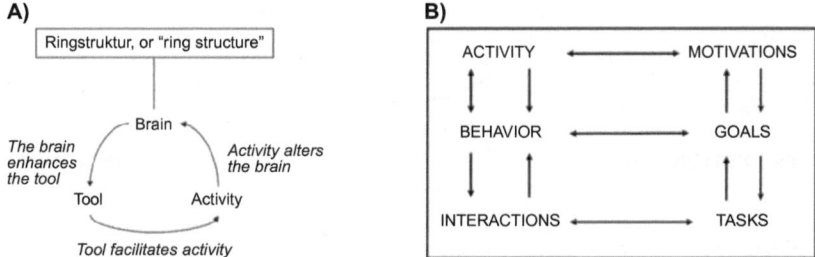

Figure 1.1 Activity theory
Activity theory is a theoretical framework for studying human practices as dynamic processes in continuous development. (A) The relationship among tool-activity-brain forms a loop resembling a "ring structure" where each component feeds back into the others. The type of tool determines the kind of activity that can be carried out, and it is this type of activity (behavior) that brings about cerebral changes, optimizing the use of the tool: "The object which the paleolithic tool-maker holds in her hand affects her mental representations (her plan, her goal) as much as those representations affect the changing object. Reciprocal relationships prevail" (Morf & Weber 2000). (B) This illustration is an adaptation of Leontiev's model to the specificities of digital behavior. The term "actions" has been replaced with "behaviors" and "operations" has been substituted for "interactions." For example, the activity of eating would be motivated by a caloric deficit. The behaviors one could engage in might vary, such as hunting, buying food, cooking it, or ordering it through a food delivery app. All of these behaviors aim to acquire food. To perform the digital behavior of ordering food through an app, numerous interactions with the technological object would be necessary, such as tapping the app icon, searching for the desired food option, and selecting the food choice, which are configured as sequential tasks aimed at obtaining the food (Hasan & Kazlauskas 2014; Leont'ev 1978).

theory proposes that the brain is an adaptive system that is constantly generating and updating its models of the world (Bayesian brain). In other words, the brain generates predictions of the sensory information that will enter the system (anticipating incoming information without knowing what it will be), and once accessed, the actual sensory information is compared with the anticipated sensory information. If they do not match, the brain corrects the prediction error in such a way that it adapts and changes its internal representations of the world to also align with the new sensory information that has entered (see Figure 1.2). This point is critically import-ant in the design of digital behavior, as the designer must consider this

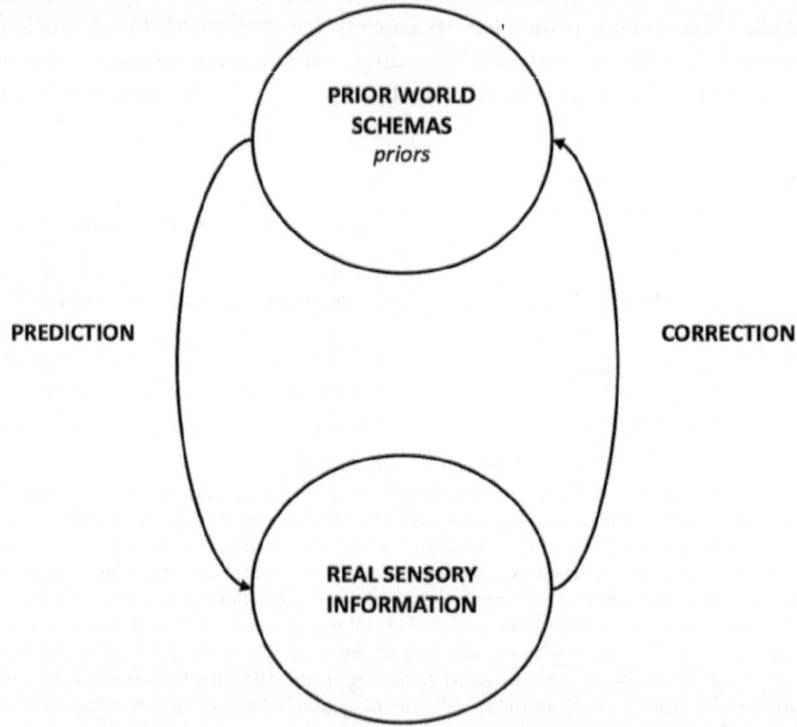

Figure 1.2 Theory of predictive coding

The brain possesses prior schemas about how the world operates. According to these schemas, the brain "predicts" or anticipates what is expected to occur in the external environment. This prediction is compared to the actual sensory information, which represents events that have transpired. The actual sensory information is then compared with the prediction, and if there is a discrepancy, the prior world schema is adjusted. This adjustment is what is considered "learning" in the theory of predictive coding.

anticipatory mechanism as a key piece in the expectations of what contingencies will appear after the execution of a specific digital behavior.

To anticipate user digital behavior, a tool already widely known in behavioral sciences will be utilized: functional behavior analysis. The distinction with this classic tool is that for the design of digital behavior, this approach will be adapted to incorporate certain theoretical elements from cognitivism. Cognitivism posits that mental representations modulate the relationship between the environment and behavior (Piccinini 2020; Pickens & Holland 2004). In other words, expectations are considered based on the understanding that they belong to the realm of temporality. If someone learns that stimulus "X" is always followed by stimulus "Y," in scenarios where "X" appears, the mental representation of "Y" will activate before the actual stimulus "Y" appears. This prior representation is what is considered an expectation (Reddy *et al.* 2015). In conclusion, digital behavior can be understood as a specific derivation of human behavior that occurs in a specific environment, namely the digital and/or virtual one.

1.4   Why Is It Important to Design Digital Behaviors?

Since the first commercial model of a personal computer in 1971 (Kenbak-1), the digital realm has pervaded all levels of our lives. This impact was particularly significant with the advent and development of smartphones in 1994, starting with IBM's Simon. However, the digital world underwent a transformation only when Apple launched the iPhone 2 in 2008, featuring the first App Store. This was the first time a popular commercial brand allowed third-party developers to offer digital services using a smartphone platform.[5] Naturally, Apple would take a significant commission. Regardless of how we arrived at the current social and technological landscape, it is clear that smartphones and their myriad possibilities have captivated human civilization. This influence extends beyond culture to even alter our brains, thereby transforming us as individuals. While this section is not intended to delve into smartphone usage statistics, it is advisable to provide some brief data. In 2008, Apple's marketplace, the App Store, started with 500 applications; by 2022, this number had ballooned to approximately 2.5 million. In contrast, Android

---

[5] In fact, the iPhone application market was not the first of its kind to exist. The concept was conceived by Steve Jobs following his ousting from Apple and subsequent founding of Next, a company focused on computer sales. In the early 1990s, Next developed the "Electronic AppWrapper," a commercial catalog of electronic software that allowed users to purchase third-party programs for their computers. Initially, this was a paper catalog, but it was later adapted for the web (www.paget.com).

and its Play Store had over 3.5 million apps. Between 2018 and 2022, it has been estimated that more than 194 billion apps have been downloaded from Android and Apple marketplaces (Aydin Gokgoz *et al.* 2021). Revenue generated from apps increased by 54 percent from 2016 to 2022, going from $101 billion to $156 billion in 2022 (Aydin Gokgoz *et al.* 2021). Notably, this market is dominated by a few; in 2019, 80 percent of all apps on the Apple Store and Google Play were created by 1 percent of developers (Inukollu *et al.* 2014). Downloading and using a technological service represents the most significant challenge for developers. Currently, the tool used for designing and enticing users to download and utilize mobile applications is the well-known Journey Map (Lavidge & Steiner 1961).

A Journey Map is tailored to the specific characteristics of marketplaces. The design and display of an app's information on the platform (Apple Store or Google Play) are key elements affecting app discoverability and, consequently, the number of downloads. An app's appearance among the top options in category-based searches is determined by factors such as its design, originality, media coverage, past revenue, downloads, app usage, and user retention (Engström & Forsell 2018). Most of these variables are beyond the control of developers and are known as "platform-controlled variables." Once a user has discovered a list of relevant apps in the marketplace's search page, they will assess which app to download. Their decision will be based on information such as the app's icon, name, position in the list, average rating, number of reviews, and price. Reviews and ratings play a critical role in the evaluation for download as they reflect the value that other users have assigned to the app. As will be discussed later, this value assignment is a fundamental process for decision-making. All these factors are known as "user-side variables" and are also associated with app downloads. The evaluation occurs on the app's description page, where further details are provided through user reviews, visual and verbal descriptions (either static or dynamic), and updates from the developers. The potential new user will assess if the app meets their needs and at what cost to decide whether the app is worth downloading.[6] The journey decision map concludes with the downloading of the app, while the same evaluation and decision-making process begins a new with other apps if the user chooses not to download (Aydin Gokgoz *et al.* 2021).

---

[6] In this process, rational decision-making is considered over emotional decision-making, in which the user is seeking the best option to fulfill a need. There are also users who make decisions more quickly due to other factors such as prior familiarity with the app they are seeking, or decision-making more grounded in emotional aspects (Kahneman 2011).

Once an app is downloaded, there appears to be a critical timeframe to persuade the user to continue using it, as 77 percent of active app users are lost within the first three days (Inukollu *et al.* 2014). For example, technical glitches like freezing, crashing, slow responses, battery drainage, or excessive ads can lead to poor app reviews. But if these technical issues are resolved, what determines whether a user will form the habit of using an app? What variables contribute to app loyalty, and more importantly, can we design or anticipate which variables will enhance user engagement? Numerous studies focus on important elements in engagement for digital (Edney *et al.* 2019; Schwarz *et al.* 2020) and virtual services (Leon-Dominguez 2022), particularly those based on gamification (Darejeh & Salim 2016; Welbers *et al.* 2019). However, there is scant literature on the theoretical foundations or a methodology for original design elements that cause engagement. Studying how users interact with the app in a real-world scenario is an effective solution for improving the digital product, but the reality is that there is limited time to convince users that we offer the service they need.

Engaging users can have a significant impact, not only on the user who fulfills a need but also on the economy of businesses that sponsor these apps or digital services. For instance, positive attitudes toward a specific business's app result in a higher purchase frequency and an increase in loyalty to the business (Bellman *et al.* 2011; McLean *et al.* 2020). Another factor that improves brand perception is if the apps are perceived as fun (van Noort & van Reijmersdal 2022). In summary, companies that invest in app development (despite already having a website) not only see an improvement in sales, but customers also spend more (Liu *et al.* 2022; van Heerde *et al.* 2019a) and increase their engagement with the business (Boyd *et al.* 2019; Cao *et al.* 2018; Gill *et al.* 2017a). Nevertheless, assessing the economic return on investment in services that produce engagement is highly complex because often the primary objective is not to sell. Rather, it is to facilitate interaction between business-to-client or client-to-client, thereby promoting the formation of an emotional and psychological bond between the customers and the business (Gill *et al.* 2017b; Inman & Nikolova 2017). One study estimated that apps with 100,000 users could generate a sales increase of 2.3 million dollars (van Heerde *et al.* 2019b). This study demonstrates that the greatest increase in online purchases (up to 9.5 percent) was made by individuals who are far removed from physical stores and those who had never bought online and were given access to the app.[7] The main difference in buying behavior in a store and through the Internet is the mediation of a digital tool to interact with the business.[8] Operationally speaking, digital behavior is the interaction of an organism (in this case, humans) with the environment (which is largely cultural), mediated by a digital interface (software) contained in a technological tool (hardware), with the aim of obtaining certain rewards (hedonic and/or utilitarian) that satisfy a need (resolve imbalances in the internal state) (see Figure 1.3).

At the functional level, digital behavior occurs because the organism seeks to restore some form of physiological (e.g., hunger), psychological (e.g., self-esteem), or social (e.g., status) imbalance through interaction with various solutions provided by digital tools. For instance, when we are hungry, we order from Uber Eats; when we wish to improve our self-esteem, we use Strava or Strong; and if we want to enhance our social status, we upload a video to Instagram, YouTube, or TikTok. These needs, which we now address through technology, have always existed and are now fulfilled through the use of digital technology (see Figure 1.4).

At this point, it is crucial to emphasize that many of the new terms frequently appearing in marketing and other disciplines related to app development are nothing more than substitutes for concepts extensively understood and studied in psychology. Terms like Growth Hacking, Customer Success, or Customer Effort Score are merely some examples that offer a biased and partial view of much more complex psychological concepts. The issue is not that these concepts are taught or that they are given much more dazzling names, but that the underlying basic psychological processes are not being taught. Consequently, they are inflexible and not generalizable to other contexts. If we do not understand the "why," we cannot adequately adapt

---

[7] Customers loyal to a brand who tend to make purchases in physical stores may exhibit a physical commitment to the sellers. However, by not utilizing the brand's application, they lack digital engagement. Therefore, it appears that as customers access the digital store and begin to interact with its features, such as reading online reviews, receiving recommendations, and easily comparing alternative products, they develop digital engagement. This digital engagement translates into an increase in purchases made through digital stores, which may even lead customers to abandon the physical store (van Heerde *et al.* 2019b).

[8] The term used is "digital behavior," rather than "technological behavior," to emphasize the fact that the interaction is not limited to the physical properties of the hardware, but also includes the digital interface that provides the organism with the necessary symbolism to interact efficiently with the digital service.

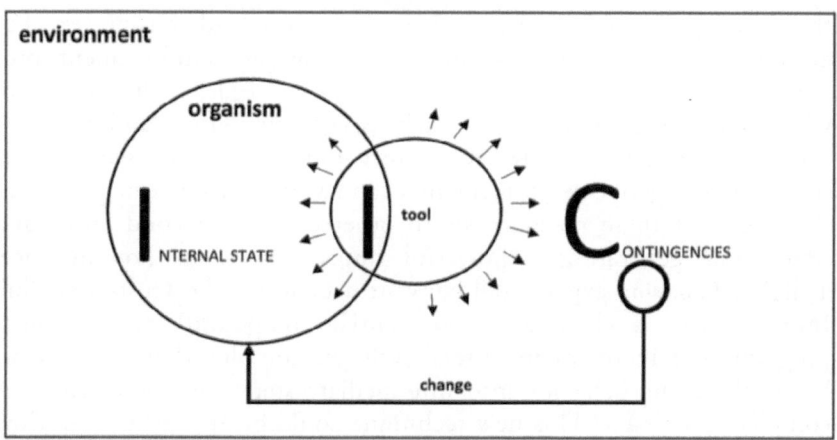

Figure 1.3 I-I-C Model
The passage of time and the environment can generate biochemical and psychological imbalances in an individual's internal state, which require addressing. Often, the solution to these needs involves the use of technological tools with which the subject interacts and produces environmental changes. These environmental changes, contingent upon technological or digital behavior, can also provoke changes in the individual's internal state. In this model, the individual who uses technology to solve a problem is considered a dynamic open system that experiences changes in response to the implementation of digital behaviors.

NEWSPAPER OF
MADRID

Sunday   October 30, 1971

Next Tuesday, a coach leaves for the City of León carrying two respectable persons; if there are another two who wish to take a seat in it for the same town or its route, they should come to the door of the Sun store of Mr. Hilario Fernández Vallejo, next to the sword shop, who will provide information on where to discuss this matter.

Figure 1.4 A representation of an authentic newspaper advertisement from 1791 in Madrid, Spain
This advertisement illustrates how many needs are socially originated and that technologies merely act as intermediaries for their fulfillment. The text has been translated into English.

them to the specific characteristics of users and businesses. For example, Customer Success consists of ongoing reinforcement programs, and Customer Effort Score is a type of negative reinforcement, which removes aversive events such as cognitive effort and the time a customer usually requires to make a purchase. Those who understand what negative reinforcement is and why it works can generalize the effects of these mechanisms to other contexts beyond effort and time, thus generating a powerful impact on sales. For instance, a digital behavior expert could invent a concept like Customer Skill Improvement, a chained (or sequential compound) reinforcement program that increases my users' skills in complex digital behaviors by breaking the behavior into intermediate steps and reinforcing the completion of each. This new technique could be applied to acquiring complex skills, such as using programs like AutoCAD or Garageband. Many nonprofessional users may give up due to its complexity, but if a company creates a Customer Skill Improvement program, it could enhance the engagement of a large portion of those users who want to learn but give up due to its difficulty. Therefore, understanding the behavioral and cognitive principles that operate on certain behaviors should be a technical imperative, not only to create more ergonomic digital services for users, but also to avoid certain actions by companies pressed by the need to produce profits when they ingeniously create reward management systems that "unintentionally" may encourage pathological behaviors in the use of technologies. An example is companies that pay people to directly play various video games. Such promotions aim "innocently" to create a community of players for certain video games, attracting other players as this artificially created community can generate interactions with new players and form emotional bonds. In this way, they manage to exponentially increase the number of players. Although this system pursues a corporate purpose, it may foster compulsive behavior to certain video games (Feiner 2023; Flayelle *et al.* 2023). So much so that already certain massively consumed and successful platforms like Instagram are implementing new features like "Take a Break." This Instagram feature aims to protect adolescents from compulsive use of the platform by setting time limits and offering alternative behaviors, such as listening to a favorite song or writing about what they are thinking, among other options. It is possible that such new functionalities arise from the multiple demands that are occurring on these types of applications, which implement behavioral principles of psychology unchecked (Crosscut 2020). Therefore, the subspecialty of Digital Behavior Design as a branch of Human Factors and Engineering Psychology, and framed within the Behavioral and Cognitive Sciences, emerges as a social need for ergonomic design of technological and digital tools. The professionalization of this sub-specialty must develop from the field of psychology, although other professionals may also be interested, such as user experience designers, interface designers, graphic designers, product engineers, and computer engineers.

CHAPTER 2

In Pursuit of Pleasure

2.1 Pleasure As a Guide to Behavior

According to Epicurus, a philosopher of Classical Greece, achieving happiness entails adhering to our hedonic nature, which revolves around the acquisition of sensory pleasure and the avoidance of suffering. Both the attainment of pleasure and the avoidance of suffering lead to pleasure; the former through joy and the latter through relief. This philosophy places hedonism as the fundamental principle that guides human behavior and has been conceptualized into natural laws by twentieth- and twenty-first-century psychology.[1] One of the earliest scholars to study how pleasure influences behavior was Edward Lee Thorndike (1874–1949), a renowned American psychologist and educator who pioneered the field of learning psychology and the formalization of behavioral principles that govern our conduct. Thorndike studied at Wesleyan University and completed his education at Harvard and Columbia. At Columbia, he earned his doctorate in 1989 under the supervision of the eminent James McKeen Cattell, and before retiring, briefly held the chair once occupied by the father of American psychology, William James. In the realm of psychology, Thorndike is primarily recognized for laying the foundation of instrumental learning, in addition to conducting the first twin studies (1905), thereby demonstrating the role of genetics in behavior and cognitive abilities (Thorndike 1905). Concerning the subject matter of this book, Thorndike was the first psychologist to postulate scientifically how environmental stimuli (S) are associated with an instrumental response (R). That is to say, stimuli present in the environment while we behave tend to form associations with the behavior being performed (Reddy *et al.* 2015). These stimuli, once associated with behavior, will in the future have a relative power to trigger the same behavior under specific contextual conditions.

---

[1] This pursuit of pleasant sensations may be genetically determined (Floccia *et al.* 1997).

Essentially, behavior depends on environmental contingencies and their effects on the organism, which will increase or decrease the likelihood of the behavior recurring in similar contexts. For example, a student who studies diligently for an exam and receives a high grade will be more likely to replicate that successful behavior – studying intensively – in future instances requiring good grades. The receipt of pleasant or unpleasant outcomes following a behavior is a central component of Thorndike's connectionist theory and, most notably, in the formulation of his famous Law of Effect. This law posits that the S-R association is strengthened when followed by a pleasing outcome, and weakened when followed by an unpleasant outcome. As Thorndike stated, "Of several responses made to the same situation, those which are accompanied or closely followed by satisfaction to the animal will, other things being equal, be more firmly connected with the situation, so that, when it recurs, they will be more likely to recur; those which are accompamed or closely followed by discomfort to the animal will, other things being equal, have their connections with that situation weakened, so that, when it recurs, they will be less likely to occur. The greater the satisfaction or discomfort, the greater the strengthening or weakening of the bond" (Thorndike 1911, 24). There is even the concept of the trans-situationality of the law of effect, which suggests that a stimulus found to be pleasant for associating with one response will likely be pleasant for associating with other responses as well (Staddon 2003). For instance, when parents use access to video games as an incentive for their child to complete schoolwork, make their bed, or clear the table, they are utilizing the trans-situationality of the stimulus. That is, playing video games holds hedonic value for the child, thereby influencing their behavior.

An intriguing aspect derived from the Law of Effect, particularly relevant to the design of digital behaviors, is the conceptualization of habit – the golden goal for any designer and developer of digital and virtual services. Habit strength relies on the associative force between the environmental stimulus and the response, to the extent that the mere presence of a cue, regardless of the outcome, could trigger the response (Wood & Neal 2007). For example, when someone starts smoking, despite the initial discomfort of inhaling smoke, the pleasurable effects of socialization can overshadow it. As a result, the act of lighting a cigarette becomes associated with certain environmental cues like bars, streets, or workplaces; over time and with repeated smoking behavior, these cues alone become sufficient to trigger the smoking action, even in the absence of the pleasurable aspects of socialization. The same occurs with technology use. Using a mobile phone typically accompanies pleasant sensations, whether we're checking social media, WhatsApp, or email. Initially, these pleasant outcomes were signaled by a notification,[2] occurring in various contexts such as work, dining, or bedroom settings. Thus, according to connectionism, an association between the context and the behavior of checking the phone was established due to the Law of Effect (Hilgard 1948). If this happens frequently enough,[3] mere exposure to stimuli present in previous contexts can prompt us to look at the phone even when there's no notification alerting us to a pleasant outcome. In other words, we automatically check our phones out of habit, since the contextual stimulus has been associated with the behavior of looking at the phone in the past, due to its accompaniment by pleasurable outcomes.[4] This is the fundamental reason why the design of notifications is so critical in the development of digital services, as they are the threads that weave habits.

The consequences following instrumental responses (by which we mean behaviors that are followed by outcomes), particularly those that are pleasurable to us, aid in better memorizing the situations and events where we can obtain pleasurable outcomes or stimuli in the future. The associations between all environmental stimuli and the response form a mental representation of the situation or event, which will later be used to identify those scenarios in which performing a specific behavior yields pleasurable outcomes (Deuker et al. 2013; Schlichtinga & Prestona 2014). In this regard, it is estimated that only a small proportion of the events we experience during the day are stored, so if the brain is to filter what to store, it would make sense for it to retain only those events that hold higher value for the organism (those events where pleasurable outcomes can be obtained in the future). A group of researchers from the Center for Neuroscience at the University of California designed an intriguing experiment in which various individuals performed a task while their brain activity was assessed through functional magnetic resonance imaging (Gruber et al. 2016). The task involved presenting the participants with visual scenes in different blocks. During the presentation of each scene, a question was posed to the participants about the scene they were viewing. If they answered correctly, they received a random reward of either two dollars or two cents. At the

---

[2] Notifications, in addition to indicating the possibility of obtaining a consequence, can be regarded as a consequence in itself when checking the mobile, as they suggest that someone is paying attention to the user (Piazza et al. 1999). Attention is a powerful reinforcer that underlies many erratic behaviors and difficult-to-explain actions, such as those of "trolls" on the Internet (Kendall n.d.).

[3] It has been proposed that the time it may take for a user to develop a habit ranges between 18 and 254 days, with an average of around 66 days (Lally et al. 2010).

[4] It is crucial to note that the contingency between behavior and its consequences has been seriously identified as responsible for habit formation in addictions (Belin et al. 2009).

end of the block, they were informed of the total amount of money won. The visual scenes consisted of a background and an overlaid figure unrelated to the background. The same background was maintained for all scenes in each block, while the overlaid image changed in each scene. For example, the questions asked were along the lines of: "Does this object weigh more than a basketball?" and a scene would appear with a basketball court background and an overlaid image of a penguin; or "would this object float?" with a background of a swimming pool and an overlaid image of a paper boat. After responding to all the scenes in each block, participants were given a rest period. However, without prior notice, they were administered a recognition memory test. In this test, only the superimposed objects (without the background) that had appeared in the experiment were presented, along with 120 new objects that had not been previously shown. The objective of the test was to recognize whether the object had previously appeared and to associate it with the background on which it was presented. The results showed that (1) the figure-background pairs best remembered by participants were those associated with a higher monetary reward, and (2) the subjects who recognized the most figures were those who exhibited greater connectivity between dopaminergic structures (substantia nigra/ventral tegmental area and the hippocampus) during the rest period following the recognition task (Gruber *et al.* 2016). In other words, as one of the authors of the study commented, "rewards help you remember things, because you want future rewards" (Davis 2016). It is the brain that is responsible for encoding the value and location of the reward in internal representations that guide behavior (Knudsen & Wallis 2021). The activity of the dopaminergic system is key to understanding motivated behavior in this sense, as it is capable of both generating various response options associated with specific consequences through mental representations (Ang *et al.* 2018); and of signaling the most appropriate option through the amount of dopamine released when considering all possible options or mental representations (Eshel 2016). In essence, we possess a hedonic brain that is designed to identify and store events with gratifying value, as we wish to re-experience the pleasures previously encountered in the past.

2.2   How Pleasure Controls Our Behavior

As we have observed, the outcomes of our actions appear to be a fundamental aspect in explaining why we behave the way we do. Indeed, the utilization of digital services should be understood as a tool that enables us to rapidly and effortlessly attain pleasure.[5] In hunter-gatherer societies, hunting could take hours or even days, whereas in modern societies, one can go to the supermarket within an hour or have food delivered to the doorstep in thirty to forty minutes through technology. In this regard, the digital behavior of ordering food via an app achieves the pleasure of eating without the need to hunt for it. If the delivered food meets expectations, the likelihood of using the app again increases, a process known as positive reinforcement. However, is receiving the food the only reason for using the app again? No. As much as you receive something you like, you also avoid things that might cause discomfort – such as having to commute, potentially interact with others, or interrupt your home activities, among other things. The avoidance of unpleasant events is also a source of pleasure and increases the likelihood of using the app again under similar circumstances, a process known as negative reinforcement. Therefore, the satisfaction gained from obtaining appetitive stimuli and avoiding aversive stimuli increases the probability of reusing the digital service or technological tool under appropriate circumstances. Essentially, the use of digital services is governed by the satisfaction derived from obtaining appetitive stimuli (e.g., food) and avoiding certain aversive stimuli (e.g., effort). The appetitive stimuli we acquire through digital tools and how they meet our expectations determine when and how these tools will be used again. Behavioral sciences have already addressed these issues through experimental analysis of behavior, applying it across various fields using a technique known as Functional Analysis of Behavior. This technique posits that behavior results from a three-term functional relationship (three sequential moments in time) involving: (1) stimuli signaling the possible presence of rewards or punishments, (2) behavior,[6] and (3) consequences in the form of stimuli. The appearance of certain stimuli that have a sensory impact on individuals serves a discriminatory function (indicating what response should be produced to obtain rewards or avoid punishments) or a controlling

---

[5] Zygmunt Bauman, a Polish philosopher and sociologist, posited that contemporary society pursues a continual sense of becoming through the instant gratification of its desires (Bauman 2000). Possibly, the advent of mobile phones and the aversion to cognitive effort (David et al. n.d.), has fostered in newer generations a psychological discomfort if what is desired is not obtained in the short term (Tangney et al. 2004). Moreover, it is conceivable that the use of this technology is altering our cognitive function (Barr et al. 2015; León-Domínguez 2024; Wilmer et al. 2017).

[6] Throughout the text, we will interchange the terms "behavior," "response," and "action" to enhance reading fluency. These three terms are similar with slight nuances, such as "behavior" being more commonly used in educational settings, "response" in experimental settings, and "action" in popular contexts.

function (indicating the success or failure of the behavior to obtain rewards or avoid punishments). This three-term relationship enables us to learn new behaviors, maintain existing ones, or even extinguish those that are no longer useful for achieving pleasure. Operant conditioning is the learning mechanism that explains the relationship between all these functional elements.

2.3 Theoretical Principles of Operant Conditioning

Operant conditioning is the term coined by Harvard Doctor Burrhus Frederic Skinner to explain how behavior is controlled by its consequences (Skinner 1938). This is an associative learning mechanism based on the functional relationship that maintains an individual's voluntary behavior with the sensory elements of their environment (stimuli). Practically, there is very little difference between this and its predecessor term, instrumental learning, and the terms are often used interchangeably in the literature.[7] Skinner developed a theoretical framework that expanded upon Thorndike's Law of Effect, centered on the probabilistic relationship (contingency) between behavior and the "appearance" of certain consequences. This is a critical point, as a key task for digital behavior designers would be to identify, design, and administer appropriate stimuli for the right context. Moreover, the timely and appropriate identification and administration of stimuli do not ensure that the user will repeat the behavior, as it is influenced by many other uncontrolled environmental and internal factors. Nevertheless, the aim of digital behavior designers is not to get all users to reuse the digital service but to increase the likelihood of reuse. Consider the following case: Amazon had revenues of 163 billion dollars from its online stores in 2019–20. It now hires a digital behavior

---

[7] In lieu of directly addressing the question of why reinforcers are effective, Skinner chose to carefully observe, define, and manipulate the situations in which certain stimuli increase the likelihood of a specific behavior. Furthermore, he provided a detailed description of the behavioral pattern that ensues from these environmental changes. In this regard, Skinner diverged from Thorndike, as the latter believed that consequential stimuli had reinforcing properties due to their ability to produce pleasure or pain. Watson rejected this terminology of "pleasure" or "pain" and, consequently, also dismissed Thorndike's Law of Effect, as these are subjective terms. Presently, it is understood that the hedonic characteristics of stimuli that produce pleasure or well-being arise because certain sensory elements interact with our sensory receptors, possessing enough power to generate an electrical nerve impulse to the brain capable of encoding a message for the limbic system. The limbic system would be responsible for activating cognitive representations (with associated neurotransmitters) and motor patterns for approach or avoidance to the stimulus. This discovery was the reason David Julius and Ardem Patapoutian were awarded the Nobel Prize in Medicine in 2021 (Reeh & Fischer 2022). Consequently, Thorndike's perspective might have been accurate.

design firm to enhance customer interactions and purchases. The firm introduces new changes, and Amazon reports only a 1 percent increase in revenue. This seemingly small figure represents a substantial amount of money at the financial level: 1 percent of 163 billion is 1.63 billion dollars. Therefore, digital behavior designers should not be obsessed with converting 100 percent of the users but should focus on enhancing the interactions of technological services with users by anticipating, signaling, and rewarding target digital behaviors.

Before delving into the detailed examination of B. F. Skinner's experimental analysis of behavior, it is essential to offer a brief introduction to who he was, given that his propositions form the central axis of this book, although they are not the only ones. Burrhus Frederic Skinner was born on March 20, 1904, in Susquehanna County, Pennsylvania. Skinner graduated with an English degree from Hamilton College in New York and displayed a strong passion and interest in becoming a writer. Despite his efforts, he did not achieve the success he had hoped for, leading to a period of depression. However, after reading Ivan Pavlov's book on "Conditioned Reflexes," along with the works of Watson and Russell, his spirits seemed to lift. This newfound motivation led him to enroll in Harvard's School of Psychology, where he graduated in 1930 and earned his Ph.D. in 1931. In the same year that his daughter Julie was born (1938), the result of his marriage to Yvonne Blue, Skinner published "The Behavior of Organisms," which, although initially not very successful and heavily criticized, has become one of the seminal works in psychology and behavioral sciences. The main critique the book received was that it treated human beings as mere machines subject to the unfolding of events, reducing them to a set of rules and stripping them of free will. At a time when American psychology was dominated by William James' introspection, which led to methodological issues, Skinner believed that psychology could be an objective and rigorous science, much like any other branch of the natural sciences. Inspired by Pavlov's account of reflexive responses in organisms, Skinner thought that an organism's voluntary behavior could be explained by objective rules, devoid of any need to understand the organism's internal states.[8] Skinner's objective was to bring order to behavioral chaos. He denied

---

[8] Skinner's approach, denying an internal causality of behavior, is quite controversial today, as it has been shown that internal representations mediate the functional relationship between the environment and the behavior of organisms (Holland & Rescorla 1975). Nevertheless, the theoretical foundation proposed by Skinner was very accurate in its predictions, even if they were ultimately not based on true statements. This often occurs in science, with possibly its most famous example being Isaac Newton's theory of gravity, which provided a very precise calculation of the movement of

the existence of randomness in animal behavior and was interested in understanding why an animal exhibits a particular behavior when it has the option not to, or to exhibit another behavior. He was influenced by numerous theories in formulating the premises of operant conditioning. On one hand, there was Thorndike's Law of Effect, which posited that the consequences of behavior create a strong connection between stimuli present during that behavior; from Watson, he adopted the use of observational techniques for obtaining objective measures. Although Thorndike and Watson were clear influences, they were only mentioned two and five times, respectively, in Skinner's most significant work "The Behavior of Organisms," while Pavlov was cited countless times. Another scholar he drew upon was Ernst Mach. From Mach, Skinner adopted phenomenalism and the functional analysis of sensations, which would later become the precursor to the Functional Analysis of Human Behavior.[9] For Mach, all knowledge originates from sensations and experience, triggered by physical stimuli. Mach also rejected the mechanistic conception of cause, focusing instead on the analysis of interdependent relationships between physical stimuli and individual experience (Skinner 1970).

Skinner adopted Mach's stance as a methodological principle but eliminated any form of internal explanation for behavior, replacing it with a functional relationship between the environment and the subject's behavior. This perspective led him to argue that the behavior of any human or animal can be explained by their past learning history, rather than internal representations of possible futures. This thought then shifted toward Darwin's environmental determinism, avoiding Mach's solipsism, and asserting that an organism's behavior is determined by environmental contingencies. In some ways, it is the environment that selects a behavior as appropriate and efficient through contingent stimuli that have some value to the organism. For Skinner, the environment is always the independent variable, and behavior is the dependent variable. Therefore, an organism's response occurs as a function of the environment and the organism's prior history in that environment: $R=f(S)$,[10] where the stimulus is any contextual event that provokes a sensory impression in the organism. This approach

the planets, even though its theoretical postulates were not true. Today, Albert Einstein's theory of relativity is considered the true theory of gravity, until it is refuted.

[9] Mach had a significant influence on major scientific currents of his time. In addition to influencing behaviorism through Skinner, he also impacted psychoanalysis and gestalt. Notably, Einstein acknowledged the influence Mach had on him, viewing Mach's philosophy as a precursor to his theory of relativity due to the critique and questioning of Newtonian mechanics (Flor 2019).

[10] By environment, we understand the set of stimuli present before, during, and after emitting a response.

led to conflicts with his psychological peers, as it eliminated all human autonomy in decision-making, which according to Skinner, was determined by environmental conditions (Skinner *et al.* 2006).

2.4 Experimental Analysis of Behavior

The methodology of experimental analysis of behavior, pursued by Skinner, allowed for the inference of the principles of operant conditioning that underlie both human and animal behavior. For Skinner, the objective of behavior study was to be able to predict it. Behavior could be predicted as long as its past history of reinforcement was known, rejecting the necessity to understand the mental or intentional states of the individual. In this regard, operant conditioning describes how the acquisition or modification of voluntary behaviors, which include digital behaviors, is dependent on the stimuli that follow them. For the laboratory study of behavior, Skinner

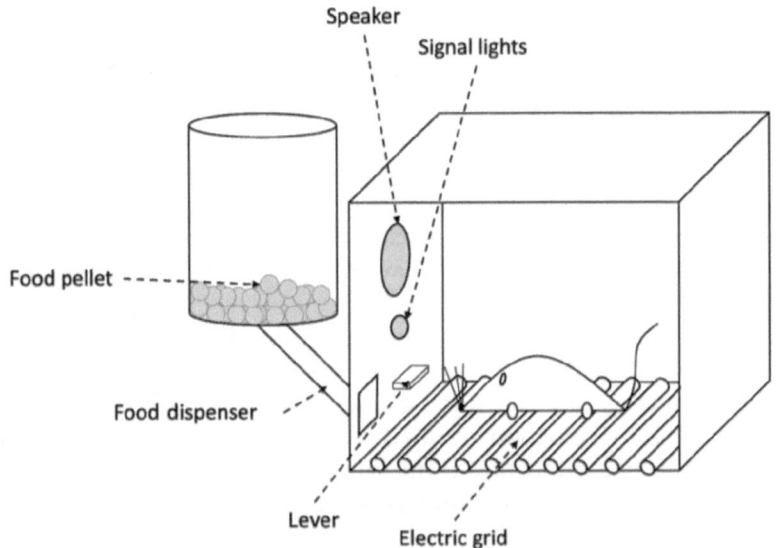

Figure 2.1 A representation of a Skinner Box
The Skinner box consists of a space with four walls shaped like a room in which the animal is placed and subjected to a learning process known as operant conditioning. This box includes a discriminative stimulus, which is usually visual and/or auditory, a tool for performing a behavior (a lever or button), and a dispenser of appetitive stimuli, such as a food dispenser. In some variants of the box, it may also feature an electrified floor to administer electric shocks as forms of aversive stimuli.

developed the renowned Skinner Box, also known as the "Standard Experimental Chamber for Operant Conditioning." This experimental chamber consists of a rectangular volumetric space where animals (rats or pigeons) could be introduced, and are administered a reward or punishment if they perform a specific behavior at a particular time (see Figure 2.1). To obtain their reward, the rat must press a lever, whereas the pigeon must peck a square. Additionally, these experimental boxes could be equipped with a speaker or lights that the animals could associate with the behavior and its consequences; or with a floor that served as an electrified grid to punish the animals with minor shocks. The presence of a visual or auditory cue could indicate to the animal that if it performs the specified behavior, it could either obtain a reward or avoid a punishment, consequences that are of hedonic nature.

Through experimental processes in these boxes, Skinner identified the fundamental principles of operant conditioning, which are considered the main framework for the design of digital behaviors. The use of the Skinner Box as the primary instrument in studies enabled the discovery of how we associate the presence of certain environmental stimuli with the opportunity to receive rewards or avoid punishments if we engage in appropriate behavior. He termed these stimuli as discriminative stimuli. Skinner referred to the rewards and punishments administered in relation to specific behavior as Contingent or Consequent Stimuli. If they increase the likelihood that the behavior will be repeated, they are considered reinforcers,[11] if they decrease that likelihood, they are considered punishments. For example, in a rat

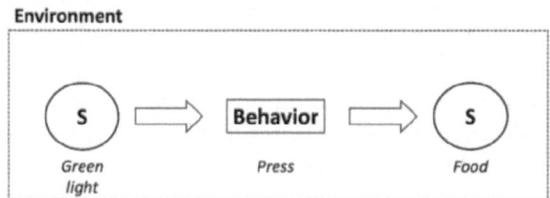

Figure 2.2 Schema of a relationship with three terms

Diagrammatic representation of a three-term relationship in which an environmental stimulus (e.g., green light) signals that contingent upon the behavior of pressing a lever, a stimulus of value to the organism (e.g., food) may be presented. This occurs because, in the past, when the organism engaged in this behavior, food was likely to appear.

---

[11] In an intriguing experiment, researchers facilitated bursting responses of individual CA1 pyramidal neurons with local micropressure applications of dopaminergic drugs. The authors proposed that they had placed a neuron inside a Skinner box, and that its dopamine-driven activity could be regarded as the "behavioral atoms" postulated by Skinner (Stein *et al.* 1994).

already familiar with the experimental process of the Skinner Box, when stimulated with a green light (discriminative stimulus), it would press the lever to obtain food (behavior), because it had previously received food (reinforcer) when pressing the lever with the light on (see Figure 2.2). One of Skinner's most classic assertions was that the rat did not press the lever because it knew food would appear, but rather because, in the past, the behavior of pressing the lever was contingently and contiguously followed by food.

The first time the rat pressed the lever with the light on was random, but with the acquisition of the reinforcer, it associated the environmental stimuli and the lever-pressing response with the reward.[12] Driven by the brain's desire for more rewards, the rat would again attempt a series of random behaviors until it once more pressed the lever and received the reward. Gradually, the rat discriminated among all the random behaviors it was performing to identify which were effective, until only the behavior of pressing the lever remained. This behavior led to obtaining food when the green light was on. Thus, the rat associated the green light with the possibility of receiving a reward. In some sense, the discriminative stimulus (green light) allows the rat to discern among the entire range of possible behaviors which is the appropriate one for obtaining the reinforcer (pressing the lever). In this case, the behavior of pressing the lever has been controlled by environmental stimuli. The control exerted by the green light, or discriminative stimulus, over the behavior is known as its discriminative function, which indicates among all possible behaviors, which would be the correct one for obtaining food. Conversely, the food serves what is known as a feedback function on the behavior, selecting that pressing the lever is the appropriate behavior to perform in a given context (when the light is green). This feedback loop is a mechanism that represents the functional relationship between environment and behavior (see Figure 2.3).

An operant conditioning involving individuals can be exemplified by a scenario where parents agree with their child that they can play with the tablet only after completing their homework (see Figure 2.4). In this example, the discriminative stimulus (S^D) might be finishing a meal, indicating that if they complete their assignment (R; operant response), they will be allowed to play Roblox on the tablet (S^C; contingence stimulus). If this

---

[12] In the specific case of rat conditioning within a Skinner Box, the initial behavior of pressing the lever does not solely arise due to the variability and randomness of their movements. The experimenter premanipulates the rats to facilitate this behavior. Firstly, the rat is intentionally placed in the box while hungry, thereby enhancing exploratory behaviors. Furthermore, they have been previously conditioned to move near the lever through a self-shaping process.

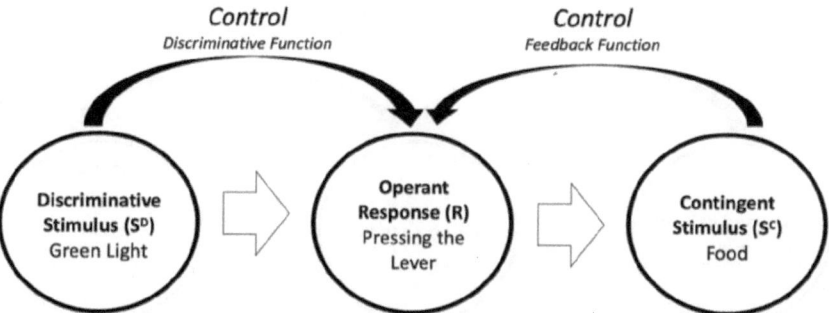

Figure 2.3 Example 1 of a relationship with three terms
The stimuli that precede behavior are termed discriminatory stimuli or deltas (depending on the pleasurable or aversive value of the stimulus they anticipate), which serve
a discriminatory function in the organism. These stimuli signal the appropriate behavior for
obtaining a consequent stimulus, which in the case of the figure, is a reinforcer. Once
administered, the reinforcer serves a feedback function, indicating the suitability of performing that operant response in similar contexts to achieve the same outcome.

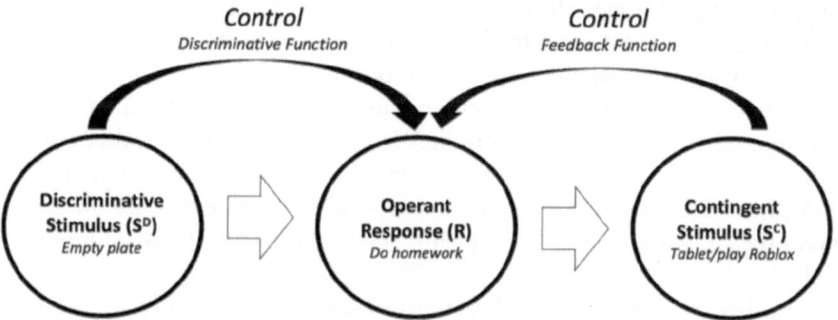

Figure 2.4 Example 2 of a relationship with three terms
This functional analysis of behavior elucidates how children have learned that upon
finishing their meal, if they complete the task, they can secure playtime on the tablet.

triple contingency "finishing a meal (empty plate)"→"completing assignment"→"playing on the tablet" is met, it may increase the likelihood in the
future that the child will prioritize completing their homework when they
wish to spend the entire afternoon playing Roblox.[13]

---

[13] In these examples, caution must be exercised when applying them, as the child may assign greater
value to playing with the tablet than to completing the task. Such situations can be examined

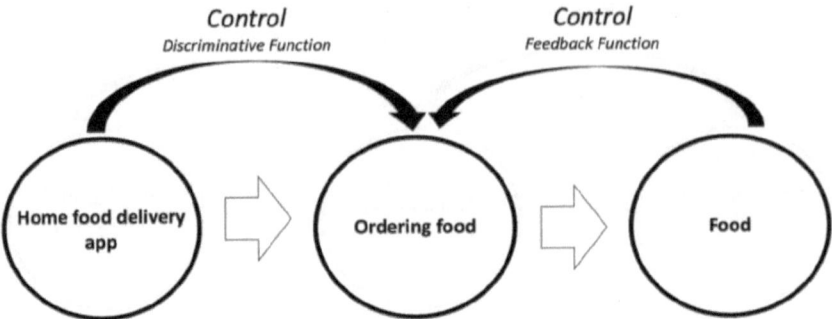

Figure 2.5 Example 3 of a relationship with three terms
This functional analysis of behavior elucidates how the digital behavior of ordering food
via app has been acquired.

At the digital level, there are exemplary instances of this functional S^D-R-S^C relationship. For instance, the mobile app icon for Uber Eats (S^D) signifies that if we order food through the app (R), we will receive food (S^C) (see Figure 2.5). Receiving the food increases the likelihood of repeating the behavior of ordering food via Uber Eats the next time someone needs food and has access to Uber Eats. Having access to the consequent stimulus is indicated by the presence of the discriminative stimulus. It is crucial to point out that hunger does not serve as a discriminative stimulus in the operant conditioning. Instead, based on operant conditioning, it is a dispositional variable that bestows value upon the consequent, and consequently, upon the discriminative stimulus as well.

Another example is WhatsApp notifications. These notifications are, in fact, signals (S^D) that indicate that if you open the application (R) you will find a new message (S^C). In this regard, we could segment operant behavior into three functionally related blocks: (1) an antecedent stimulus that signals the possibility of obtaining a reinforcement (consequence), (2) an operant response, which is the behavior performed to obtain the reinforcement, and (3) the contingent stimulus (or reinforcer) that selects the operant response as appropriate in a given context. Subsequently, a summary will be provided on the nomenclature of all terms involved in the Experimental Analysis of Behavior to facilitate the understanding of these relationships (see Figure 2.6).

through the principle of Premack, which suggests that preferred activities can be used to reinforce less pleasurable activities (Herrod *et al.* 2022).

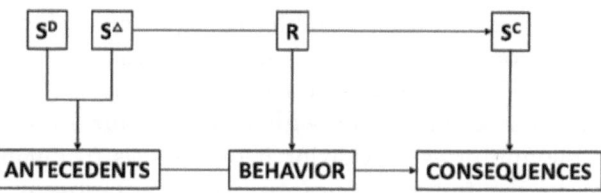

Figure 2.6 Behavioral nomenclature
discriminative stimuli and deltas are known as antecedents, as they occur prior to behavior. The operant response is referred to as behavior and action, and its contingent stimuli are called consequences.

Antecedents constitute a descriptive category that encompasses a set of stimuli present when a response is emitted, preceding the consequent stimulus. This category is composed of two types of stimuli: (1) discriminative stimulus (S^D) that signals the possibility of obtaining a pleasant consequent stimulus contingent upon behavior; and (2) delta stimulus (S^Δ) that indicates the possibility of receiving an unpleasant consequent stimulus contingent upon behavior.[14] The operant response (R) is the behavior that the organism needs to perform in order to obtain the reinforcer. Importantly, this response can be described either topographically or functionally. Skinner, to emphasize the central aspect of behavior, introduced the concept of "response class," which encompasses all the topographies (or forms) of responses that serve the same function. That is, they lead to the same reinforcer. For example, individuals can engage in different behavioral topographies to obtain a food reinforcer, such as going to the supermarket to buy it, going to a restaurant and waiting to be served, or ordering it via mobile phone for home delivery. All these operant responses belong to the same response class because they each yield food (serve the same function). On the other hand, describing the topography of the response would mean detailing the sequence of organismic movements involved in performing the behavior. Skinner described the topography of a rat's behavior when pressing a lever as follows: "The entire behavior of lifting up the fore part of the body, pressing and releasing the lever, reaching into the tray, seizing the pellet of food, withdrawing from the tray, and eating the pellet is, of course, an extremely complex act. It is a chain of reflexes, which for experimental purposes must be analyzed into its component parts" (Skinner 1938, 51).

[14] The delta stimulus can also signal an extinction process, which will not be addressed in this book. Therefore, the delta stimulus should be considered as a stimulus signaling the onset of punishment if a specific behavior is exhibited.

Finally, there are contingent stimuli (consequences), which serve as the central axis of operant behavior and are crucial for the design of digital behaviors. The term "consequent" is merely a descriptive category that encompasses various stimuli with a feedback function on behavior, indicating the appropriate behavior if one wishes to obtain the same consequent in the future.[15] Consequences or contingent stimuli are categorized into two groups: (1) those that increase the likelihood of a particular response reoccurring are termed "reinforcers," (2) whereas those that decrease this likelihood are called "punishments." Generally, a stimulus is a reinforcer if it is pleasant, and a punishment if it is unpleasant, although this is not always the case.[16] Moreover, these stimuli are also defined by their type of contingency, which can be either positive or negative. A positive contingency exists if the likelihood of the consequent appearing after the operant response is higher than the probability of it not appearing; conversely, a negative contingency exists if the likelihood of the consequent not appearing after the operant response is greater than the probability of it appearing. Broadly speaking, the processes for obtaining consequences can be defined as follows (see Figure 2.7):

- Positive Reinforcement (S^{C+}): An appetitive stimulus that emerges following the emission of the operant response (positive contingency), which increases the likelihood of the same behavior reoccurring in a similar environment (e.g., a hug after helping with a problem, a "like" on Instagram from someone you are attracted to).
- Positive Punishment (S^{C-}): An aversive stimulus that appears after the emission of the operant response (positive contingency), reducing the likelihood that the same behavior will occur again in a similar setting (e.g., a punch after insulting someone, a damaging comment on a Facebook post about a specific topic).
- Negative Reinforcement (noS^{C-}): An aversive stimulus that does not appear after the emission of the operant response (negative

---

[15] It is essential to remember that Skinner asserted that we engage in behaviors not because we seek to obtain a reinforcer, but because such behavior was reinforced in the past. In the digital behavior design model, being human and having phenomenological experiences, we are capable of envisioning a future reward because we have been rewarded in the past. This nuance is of paramount importance, as it implies that humans can generate expectations and act accordingly.

[16] Consider the case of X (formerly Twitter). Likes and retweets tend to be pleasant stimuli, but they can become unpleasant depending on who administers them. For instance, if I identify myself as a social democrat and post a tweet about the beautiful funeral of the Queen of England, and an individual known for their fascist ideals likes it, that like becomes unpleasant. It could decrease the likelihood of me commenting positively about the Queen of England or any monarchy in the future.

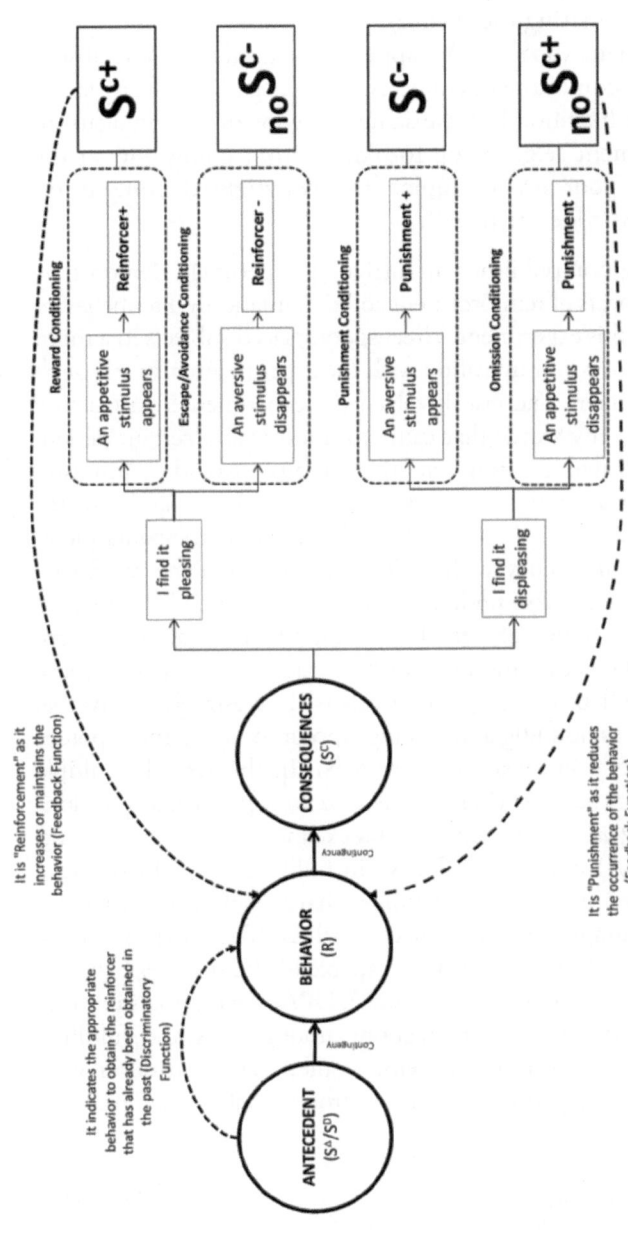

Figure 2.7 Simplified schema of a three-term relationship

The present figure provides a simplified schematic representation of the functional relationships based on three terms. The antecedents ($S^\triangle/S^D$) serve a discriminatory function on behavior (R), which is followed by a contingency that holds value for the organism (S^C). According to Thorndike's Law of Effect, the value of the stimuli contingent upon behavior can be reduced to an affective factor of either well-being (appetitive stimulus) or distress (aversive stimulus). The affective value of well-being can be generated either by a positive contingency of an appetitive stimulus, known as reward conditioning (S^{C+}; positive reinforcement), or by a negative contingency of an aversive stimulus, known as escape conditioning (if the aversive stimulus is already present) or avoidance conditioning (if the aversive stimulus is about to be presented). Both escape and avoidance conditioning operate through a process of negative reinforcement (noS^{C-}; negative reinforcement). On the other hand, when the value of the stimulus contingent upon behavior causes distress in the organism, it could be due to a positive contingency of an aversive stimulus post-behavior, known as punishment conditioning (S^{C-}; positive punishment); or if post-behavior a negative contingency of an appetitive stimulus occurs, it is known as omission conditioning (noS^{C+}; negative punishment).

contingency), and increases the likelihood of the same behavior reoccurring in a similar environment (e.g., a friend starts talking to you again when you speak positively about their partner on social media, thereby eliminating the silence).

- Negative Punishment (noS^{C+}): An appetitive stimulus that is absent following the emission of the operant response (negative contingency), which reduces the likelihood of the same behavior occurring again in a similar environment (e.g., being forbidden from going out on the weekend due to poor grades, uploading copyrighted content on YouTube leading to video removal).

In 1955, Joel Greenspoon published an intriguing experiment demonstrating how even subtle aspects of reinforcement could dramatically alter behavior (Greenspoon 1955). In this experiment, Greenspoon placed subjects in a room where their only form of communication with the experimenter was through a microphone and speaker. The task involved the experimenter instructing the subjects to vocalize any words that came to mind over a certain period. The experimenter, listening from another room, would administer a contingent verbal social reinforcer in response to plural or singular words ("mmm-hmm" or "huh-uh"). Results indicated that the participants incrementally increased the frequency with which they vocalized the reinforced words (Greenspoon 1955). These findings have been replicated in numerous studies utilizing various types of verbal and nonverbal social reinforcers, effectively increasing the vocalization of words in categories such as animals, verbs, hostile words and pronouns, among others. Interestingly, nonverbal social reinforcers, such as nodding and smiling, appear to have a more potent effect than verbal social reinforcers (Krasner 1958). In this regard, multiple social reinforcers such as gestures (Williams *et al.* 2020), positive impressions (reputation) (Izuma *et al.* 2008), smiles (Spreckelmeyer *et al.* 2009), and compliments (Deci 1971) have been effective in fulfilling individuals' social needs. In digital environments, these reinforcers have evolved into what are known as Digital Paralinguistic Affordances[17] (DPAs; likes, sharing, views, etc.) (Ait Oumeziane *et al.* 2017), which are capable of altering user behavior on social media platforms (Lindström *et al.* 2021). DPAs serve as tools in social media, facilitating communicative interactions among users and fulfilling diverse needs for the user who grants them, or they may even occur with no apparent purpose, serving as an automated process (habit). For example,

---

[17] In digital environments, "likes" and "favorites" have been conceptualized as Digital Paralinguistic Affordances (DPAs), which are stimuli that facilitate nonverbal social communication.

liking a post on social media could be related to a desire for status or simply for the sensation of feeling connected (Hayes *et al.* 2016). On social media, digital behaviors such as uploading a photo, video, opinion, or comment usually come with user feedback in the form of DPAs, which act as reinforcers or punishments by signaling approval or disapproval from another community member. These DPAs are highly significant and revealing, as they indicate real-world personal traits such as personality traits, sexual orientation, ethnicity, political and religious stances, intelligence, happiness, addictive substance use, parental status, age, and gender (Kosinski *et al.* 2013). Observing social stimuli generally increases activation in the prefrontal cortex and the striatum when the stimuli hold high value for individuals (Calabrese *et al.* 2022). Striatal activity also elevates when thinking about individuals with whom intimate relationships have been maintained (Fisher *et al.* 2006; Hughes & Beer 2012) or who are held in high esteem (Zink *et al.* 2008). That is, behaviors that trigger thoughts of high-value social stimuli can also be reinforcing, such as a photo or video on a social media platform. Intriguingly, men activate a broad mesolimbic network in response to monetary reinforcers but only a partial network for social reinforcers. Women, however, activate the same network amplitude for both types of reinforcers. This suggests that the value of social and nonsocial reinforcers could differ between gender, a consideration that should be accounted for in the design of digital services.

2.5 Examples of Operant Behaviors in Digital Environments

Following this brief overview of the foundational pillars that will guide the design of digital behavior, this section will discuss some examples of digital behaviors with which we are all familiar under the principles of operant conditioning. Here, the key point is understanding that the individual already has a prior usage history of the digital service, thereby enabling the anticipation of the appearance of reinforcers and punishments. The first example will examine the digital behavior of an individual receiving a notification from Instagram (see Figure 2.8).

This analysis is post hoc, meaning it occurs after an individual has viewed Instagram, allowing for the determination of the type of reinforcer received. In this instance, through the Instagram notification (S^D), the individual anticipated a potential reinforcer. Context here is paramount. Imagine being in a lecture. Viewing the content of the message within the Instagram application is time-consuming (positive punishment; the professor reprimands me), so viewing the content of the message directly on the notification screen could be a highly likely option (negative

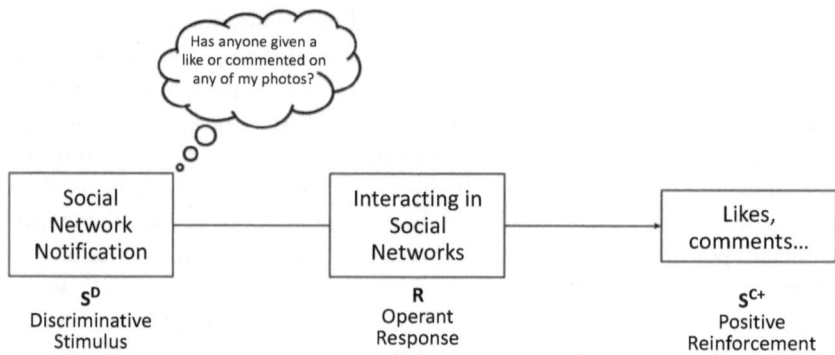

Figure 2.8 Operant conditioning in the use of the social network as Instagram
Instagram notifications serve as an antecedent stimulus signaling that if one chooses to open the app, pleasant content will be available. For this reason, social networks aim to minimize negative interactions to the greatest extent possible, so that their discriminative stimuli anticipate appetitive stimuli. The anticipation of aversive stimuli leads to avoidant behavior among users. Hence, applications like X (formerly Twitter), Instagram, or TikTok lack a dislike button; no one wants to anticipate outcomes of negative affective value.

reinforcement; it eliminates the penalty of taking too long and being reprimanded by the professor). In another scenario, suppose you are in your room, lying on the bed relaxing, leading you to decide to access the Instagram application directly. Upon doing so, you realize you've received numerous likes on your latest post. Consequently, two reward conditionings have taken place. The first conditioning pertained to the behavior of accessing Instagram upon receiving a notification, as it may indicate the presence of numerous positive reinforcers, which in this instance were likes (S^{C+}). The other reward conditioning was somewhat more nebulous or deferred (extended contiguity between behavior-contingent stimulus). The behavior of uploading a post with specific content to Instagram was conditioned, given the high number of likes or views (S^{C+}). Furthermore, if you access Instagram, other reinforcers will automatically be available in the form of posts you find intriguing, making the act of accessing Instagram conditioned by rewards from the posts appearing on your feed. These rewards can become endless with the infinite scroll feature on social media platforms. This intricate web of reinforcers causes our brain to associate Instagram with the acquisition of pleasure-inducing reinforcers through the release of dopamine (Fraser *et al.* 2023). In this manner, our hedonistic brain will associate all stimuli present before,

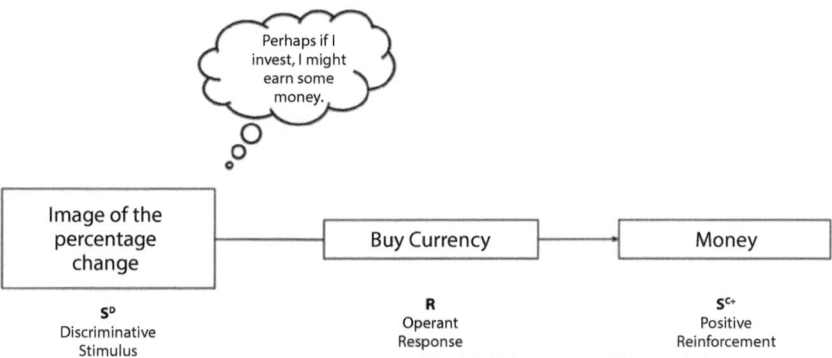

Figure 2.9 Functional analysis of a stock value app
In this example of the digital service, the appearance of a series of discriminative stimuli in the form of icons on the main screen of the app anticipates the potential acquisition of money if one buys the currency with an upward trend. This anticipation of the appetitive stimulus leads the user, through metacognitive processes, to consider the action of purchasing said currency.

during, and after obtaining the reinforcer, forming a mental representation that signals the potential to acquire pleasure-causing reinforcers.

Another example of how digital stimuli control digital behavior can be observed in mobile applications for cryptocurrencies or financials (see Figure 2.9).

The initial screen is designed with ample information regarding various indicators about the real-time economic status of your cryptocurrency investments and the market. Every icon or piece of information that appears in the app is an S^D signaling that if you engage in the behavior of pressing it, you may be directed to another screen with more detailed information valuable for increasing or securing your investments (S^{C+}). Following this logic, we will focus on a type of discriminative stimulus that appears on the main screen, the icons displaying the percentage changes in the value of cryptocurrencies compared to the previous day. These icons with percentage changes act as discriminative stimuli, indicating to the individual that these coins might be in a bullish phase, and if they invest in them soon (R), it could yield significant profits (S^{C+}). Therefore, for the proper design of digital behavior, it is crucial to understand the relationship between antecedent and consequent stimuli, as the design of an effective discriminative stimulus signaling easy access to the reinforcer is the key, albeit not the only, element for an ergonomic design of digital and virtual services.

CHAPTER 3

Digital Operant Box

3.1 Digital Services As Digital Operant Boxes

The concept of the Digital Operant Box (DOB) is a psychological construct that directly alludes to and theoretically acknowledges the Skinner Box. Through this construct, it is intended to recreate the principles of learning derived from the experimental analysis of behavior and find their analogs in the digital and virtual environment. The DOB can be any digital tool through which the user interacts sensorimotorly, with these interactions being predictable by designers of digital behaviors. Predictions about how users will use technological tools will form the foundation of the digital design of technological services. Skinner utilized a box of small dimensions to experimentally develop operant behaviors because it was a medium that could be limited and controlled (enclosed environments). Unquestionably, smartphones meet these criteria of being "a limited and controlled medium." Smartphones are the primary DOBs for designers of digital behaviors, though any other technological tool with a digital interface could also serve as one, such as ATMs. Just as in the Skinner Box, a smartphone is a controlled environment where the designer has the control to administer stimuli contingent upon the user's digital behavior. Additionally, in DOBs, it is possible to design the sensory aspects of discriminative stimuli and reinforcers, as well as the rules by which these discriminative stimuli will be displayed and the reinforcers administered. All these reasons are why the smartphone is used as a metaphor for the Skinner Box, a DOB. Unlike the Skinner Box, in a DOB, designers do not have control over the internal states with which users come to it, so anticipating these internal states will be a fundamental characteristic in the design of digital behaviors (see Chapter 5).[1] Users come to DOBs to meet physiological and psychological needs by obtaining a reinforcer, which can be administered

---

[1] In open and natural settings where users frequent to engage with digital services, programming is not feasible. The rats Skinner used in his experimental box were deprived of food so that the sustenance presented as a reinforcer would be highly motivating for them to perform the desired behaviors.

directly by the digital space (e.g., music played from the Spotify app) or indirectly (e.g., food from Uber Eats delivered to the home). The digital space is sometimes the direct provider of reinforcers, while at other times, when the main reinforcer is obtained outside of the digital space, the technological tool is merely a facilitator. In this sense, what happens in the digital space has repercussions in real life. An example of indirect reinforcers is the products purchased through Amazon's e-commerce stores. These consumer products arrive at the home in their physical form. Conversely, watching a movie through Amazon Prime can be considered a direct reinforcer, since the digital service itself administers the reinforcer. Another characteristic of the reinforcer related to its administration is the time elapsed between the digital behavior and its attainment. The acquisition of the reinforcer can be immediate (e.g., listening to music on Spotify upon pressing play, or accessing Photoshop features after paying for the membership) or delayed (e.g., purchasing a book on Amazon). In these cases, where the reinforcer is delayed, part of the business is transferred from the digital ecosystem to the physical ecosystem. Specifically, in the case of Amazon: relationships with suppliers, logistics, shipping, human resources, among others. This is a key point because one of the reasons, not the only one, that a well-designed digital business may fail, can be due to the part of the business that is outside the digital world. Having a good design for a digital product does not guarantee a successful business. There are other types of businesses whose environment is almost exclusively technological and digital, and the attainment of reinforcers is immediate to the digital behavior. This is the case with Spotify. A user of this app can listen to music immediately after the response of pressing the play button. Here, the bulk of the business lies in the design and development of new functionalities or features to continue convincing the user that they are the best app for administering a musical reinforcer.

3.2 Classification of Reinforcers in Digital Services

Immediate reinforcers of technological services are those obtained by users with a low contiguity between the behavior and the reinforcer. That is, the time that elapses between the digital behavior and the appearance of the reinforcer is zero or a few seconds. Conversely, when the reinforcer is delayed, the contiguity between behavior and reinforcer is greater than

However, in the design of digital behaviors, it is crucial to segment the user by their unresolved needs, as these will modulate their usage. For this reason, segmentation by traditional marketing socio-demographic variables takes a backseat in the design of digital behaviors.

several seconds, even days, as occurs with Amazon services. Furthermore, one can also identify direct and indirect types of reinforcers. Direct reinforcers are those offered directly by the technological service, whereas indirect reinforcers activate a system or process through which the reinforcer is obtained outside the digital medium, in physical reality. Thus, reinforcers acquired through digital means can be classified according to three characteristics: contingency (positive versus negative), contiguity (immediate versus delayed), and mediation (direct versus indirect):

- Immediate direct positive reinforcer (S^{C+id}): A stimulus contingent upon a digital behavior with a contiguity of zero or a few seconds, administered directly by the technological object where the digital behavior occurred (e.g., Spotify music when the play button is pressed).
- Immediate indirect positive reinforcer (S^{C+ii}): A stimulus contingent upon a digital behavior with a contiguity of zero or a few seconds, administered by an entity outside the technological object where the digital behavior occurred (e.g., turning on the room light through the Philips Hue app or a similar one).
- Delayed direct positive reinforcer (S^{C+dd}): A stimulus contingent upon a digital behavior with a contiguity exceeding several seconds, administered directly by the technological object where the digital behavior occurred (e.g., a like or comments on a post on a social network).
- Delayed indirect positive reinforcer (S^{C+di}): A stimulus contingent upon a digital behavior with a contiguity exceeding several seconds, administered by an entity outside the technological object where the digital behavior occurred (e.g., receiving food through Uber Eats).
- Immediate direct negative reinforcer (noS^{C-id}): A stimulus not contingent upon a digital behavior with a contiguity of zero or a few seconds, administered directly by the technological object where the digital behavior occurred (e.g., entering all the requested data in a mobile app eliminates the blockage, allowing its use. Or in Candy Crush, purchasing a life removes the block that allows continuing to play).
- Immediate indirect negative reinforcer (noS^{C-ii}): A stimulus not contingent upon a digital behavior with a contiguity of zero or a few seconds, and that is administered by an entity outside the technological object where the digital or virtual behavior occurred (e.g., allowing entry to a person waiting at the door through an app that controls the smart lock. The block preventing the person's entry is removed).
- Delayed direct negative reinforcer (noS^{C-dd}): A stimulus not contingent upon a digital behavior with a contiguity exceeding several seconds,

administered directly by the technological object where the digital behavior occurred (e.g., forgetting the password of a cryptocurrency electronic wallet and requesting its recovery. The process of removing the restriction that prevents entry into the electronic wallet takes several hours as it must be verified that the person is the account owner).

- Delayed indirect negative reinforcer (noS^{C-di}): A stimulus not contingent upon a digital behavior with a contiguity exceeding several seconds, administered by an entity outside the technological object where the digital behavior occurred (e.g., the restoration of electricity service due to non-payment. The service restoration following the payment of the debt through an app is not immediate).

In the design of digital behaviors for a new service or during a technological innovation process, it is paramount to identify all types of reinforcers that could potentially be offered, and to study the feasibility of each. Often, we may encounter reinforcers that were not considered in the initial draft of the app design. Sometimes, such a reinforcer only requires minor changes in the digital service, but at other times, the reinforcer is so influential that it necessitates a significant change in the service design.

3.3 Molar and Molecular Behavior

Historically, behavioral sciences have described two seemingly opposing perspectives to explain behavior. The first perspective views behavior as discrete units, termed "molecular behavior." Conversely, molar behavior has been the alternative view, positing that behavior is composed of behavioral patterns or activities that, by their nature, extend unitarily over time and cannot be explained as the sum of their parts. This is a debate between analyzing behavior from an atomistic stance or a holistic stance (for an in-depth review, see Baum 2004). With regard to the design of digital behaviors, digital behavior is intentionally molar (goal-directed), but it is constructed from smaller segments of behaviors (molecular) that aggregate and ultimately shape the resulting molar behavior (see Figure 3.1). That is, for the design of digital behaviors, it is recognized that the molar behavior of using a service (e.g., ordering food through Uber Eats) is the result of an aggregate set of molecular behaviors (e.g., opening the app → viewing the menu → clicking on food → Add to Cart → Purchase). Each of these molecular behaviors follows the same operating principles as if it were the molar behavior, that is, they are

MOLECULAR BEHAVIOR

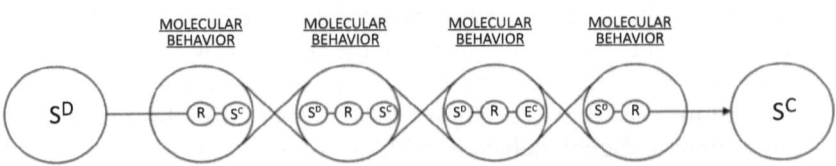

Figure 3.1 Integration of molar and molecular perspectives of behavior
Molar behavior can be segmented into various molecular behaviors. Each molecular
behavior is explained through the relationship of three terms: a discriminative stimulus
(S^D), which signals that if a specific operant response (R) is made, a determined consequent
stimulus (S^C) will follow, indicating that this is the appropriate operant response if the user
wishes to replicate the same consequences in the digital space. The S^D of the molar behavior
is the same as the first S^D of the first molecular behavior. In turn, the S^C of the molar
behavior is identical to the S^C of the last molecular behavior. Thus, molar behavior is
connected to molecular behaviors at their endpoints.

controlled by the antecedent and consequent stimuli of behavior. Thus, molecular behaviors organize the molar behavior moment to moment, while the molar behavior directs the molecular behaviors over time. This conception of behavior allows for a more detailed analysis of digital behavior (Shimp 2017).

Let's return to the case of Uber Eats to illustrate the relationship between molar and molecular behavior. The molar behavior would be "purchasing food through the Uber Eats mobile application when my refrigerator is empty" (see Figure 3.2).[2]

In an operant analysis, the discriminative stimulus signals that, if one engages in the behavior of ordering food through Uber Eats, one can obtain food. This would be a molar behavior with an indirect delayed positive reinforcer.[3] "Ordering food through Uber Eats" is simply the general name given to the set of molecular behaviors necessary to achieve the food reinforcer, which is the ultimate goal of the user. To reach this goal, the individual has had to sequentially perform multiple molecular behaviors until clicking the button to confirm the order. At this point,

---

[2] It is advisable to conduct more extensive descriptions of molar behavior to contextualize them and understand potential environmental determinants. Having information about the context can help us identify different environmental elements that may improve the design of digital behavior.

[3] Indeed, Skinner's operant conditioning suggests that the subject orders food from Uber Eats because, in the past, ordering food via Uber Eats when there was no food in the refrigerator worked to obtain sustenance.

Figure 3.2 Example of molar behavior
This diagram illustrates the molar behavior of "ordering food through a mobile food delivery app when my refrigerator is empty," where the S^D is the empty refrigerator, indicating a positive S^C for food if the operant response of ordering food via app (R) is carried out.

from the perspective of digital behavior design, it is crucial to understand that the mere sight of an empty refrigerator activates different mental representations of possible responses to obtain food such as going to the supermarket, dining at a restaurant, or visiting one's parents for a meal, among others. The selection of one response over others will be discussed later. Now, the example focuses on the moment when the decision to use Uber Eats for obtaining food has already been made (see Figure 3.3).

To successfully request food via the app, one must engage in a sequence of molecular digital behaviors, which operate under the same principles as the molar digital behavior. In the first molecular behavior, already utilizing the digital service, the discriminative stimulus is the Uber Eats app icon, which signals an immediate direct positive reinforcer, which is to see what food is available for ordering.[4] Upon performing the molecular digital behavior of clicking on the icon, another molecular segment of the molar behavior of ordering food commences. This behavioral segment begins with one or several discriminative stimuli that signal various reinforcers of differing value to the user. For instance, the icon of a rotisserie chicken chain indicates a high-value appetitive stimulus to the individual, thus, after a brief cost-benefit analysis (Spier 1971), the user clicks on the discriminative stimulus. This process continues until reaching the final discriminative stimulus, which is to confirm the order, embodying an immediate direct positive reinforcer (a notification that your order has been placed) and a delayed indirect positive reinforcer, which aligns with the molar behavior's reinforcer, receiving the chicken at home (see Figure 3.4).

Ultimately, designing a digital service is predicated on this principle of identifying the molar behavior and shaping the most suitable molecular

---

[4] It is possible to analyze the discriminative stimuli that led to the use of the Uber Eats application long before, but for this example, the analysis will begin from the moment the interaction with the digital interface starts.

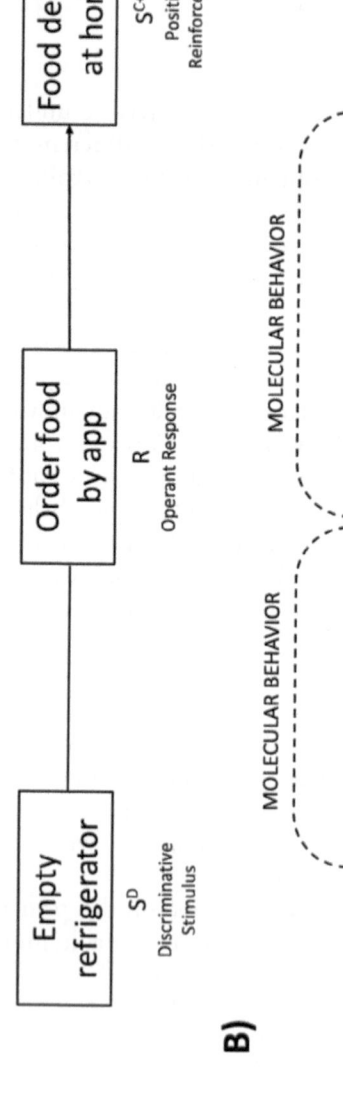

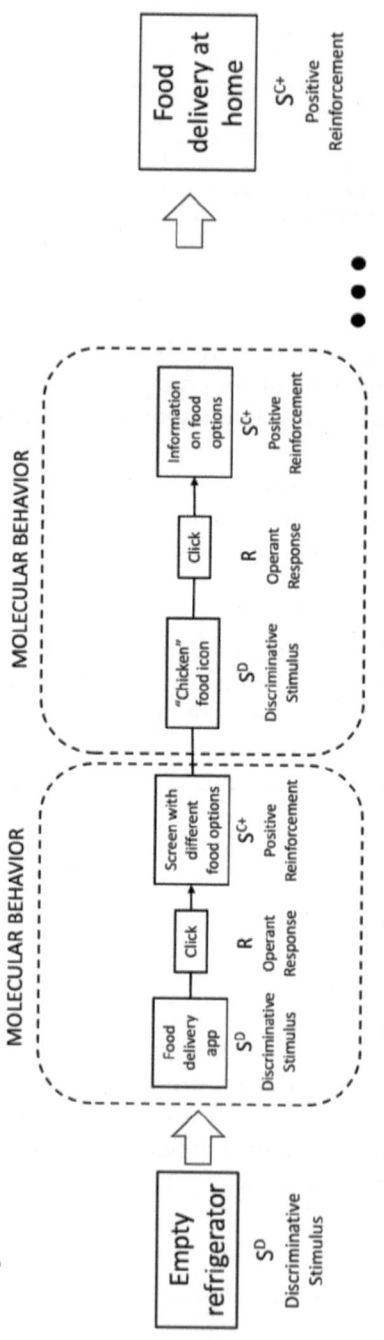

Figure 3.3 Illustration of a molar behavior segmented into molecular behaviors

A) The molar behavior of ordering food via a food delivery app is triggered by an empty refrigerator. The empty refrigerator, coupled with the accessibility of the food delivery app, allows the individual to contemplate the option of ordering home delivery food through the app. B) Once the individual decides to request food via the app, a series of molecular behavior sequences begins, which, if the service is well-designed, will culminate in the confirmation of the food order and its delivery to the home.

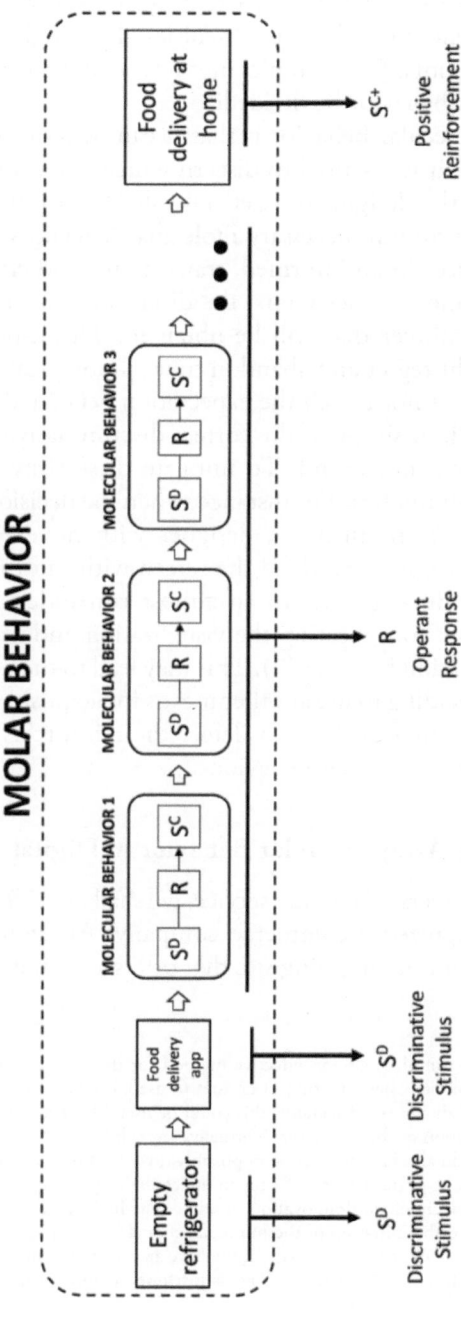

Figure 3.4 Schematic representation of the example with a food delivery app

The molar behavior of ordering takeout food through an app is comprised of different molecular behaviors that are sequentially activated until the reinforcer of the molar behavior is obtained.

behaviors to reach the molar behavior's reinforcer. Nevertheless, special attention must be paid to potential difficulties that might arise in the design of digital behaviors. For example, into how many molecular behaviors should the molar behavior be divided?

The issue with molecular behavior is that it can be atomized far more than necessary, hence it is essential to discern which behavior adds information and value to the design. As a general rule, in an initial draft, one should include the minimum necessary molecular behaviors to attain the desired molar reinforcer in an informed, transparent, and effortless manner. Moreover, in molecular behaviors, the discriminative stimulus must clearly signal the reinforcer that will be obtained. This aspect is crucial because the user might regret and abandon the technological service if the signaled reinforcer does not match the expectation generated through the discriminative stimuli, or simply if the correct discriminative stimulus for the sought reinforcer is not found. To mitigate these risks, the work of graphic designers is paramount. For instance, when the decision to click on the purchase button is imminent, a designer with no experience, but common sense, might opt to overload the screen with information about other products, introducing noise to an almost concluded purchase, in addition to potentially complicating the visualization and accessibility of the purchase confirmation button (S^D). This may lead to the user giving up on the purchase or deciding to use another means for acquisition.[5] Next, an example will be presented on how to detect the reinforcers of both the molar behavior and its molecular behaviors.

3.4 Reinforcer Array of Molar Behavior in Digital Services

To identify the reinforcers of a digital service, we shall consider an example from a globally recognized e-commerce company: Amazon. Amazon is widely known as an online shopping site that offers an extensive variety of

---

[5] Identifying discriminative stimuli is also an effective method for the technological innovation of digital services. In the example, where the subject decides to use Uber Eats to get food due to the discriminative stimulus of an empty refrigerator, this could be useful for generating an innovation that anticipates the acquisition of the reinforcer. Normally, our subject might have a cell phone in their pocket, but they would also have to retrieve it, possibly download the new version of the app, deal with a low battery, or charge the device . . . Many things could happen until they decide to make the purchase. Therefore, a technological innovation could be to reduce the time between the molar discriminative stimulus and the initiation of the molecular digital behavior sequence. How? Well, Uber Eats could sell specialized refrigerators that already have an integrated smart assistant to make grocery purchases or schedule restaurant meals. This is just one example of how digital behavior design can also be leveraged for innovation.

products, including books, electronics, clothing, food, furniture, and other items. Furthermore, it has diversified its business to include services such as Amazon Web Services (cloud services), Amazon Prime (subscription to Amazon sub-services like delivery, movies, video games, etc.), Amazon Pharmacy (online pharmacy), Amazon Prime Air (delivery by own aircraft and drones), Amazon Music Unlimited (music streaming platform), Amazon 4-Star (physical stores), Audible (podcast services), Amazon Go and Go Grocery (supermarkets), Amazon Advertising (advertising services on Amazon stores), Alexa and Echo (voice assistants), Amazon Fashion (fashion brand), Kindle and Amazon Book Store (digital bookstore), and Amazon Fresh (fast food home delivery). To exemplify the reinforcers offered by its service, we will use the case of Esther, a twenty-one-year-old student about to start a new semester in medical school. Upon reviewing the bibliography for her new subjects, Esther realizes she needs a book titled "Introduction to Pathological Physiology I" and decides to purchase it through Amazon (see Table 3.1). What reinforcers could have been involved in Esther's behavior of buying the book through Amazon?

At the top of the table, the molar digital behavior is presented, which defines the main functional relationship of the individual when interacting with the digital service. In addition to receiving the book, there is another mesh of reinforcers that sustain the behavior of purchasing books through Amazon rather than visiting a bookstore. Besides the obvious benefits, such as finding the book on the platform (eliminating the risk of going to the bookstore and not finding it) and, of course, receiving the book, these reinforcers are varied. They may include not having to use the car or interrupting activities at home (e.g., for home workers or mothers with children), or not having to interact with anyone (e.g., for those with introverted personality profiles). All these reinforcers, in some way, are bolstering the molar digital behavior of buying a book through Amazon. Conversely, one can also anticipate potential punishments such as receiving a book in poor condition. It is also of interest to identify the punishments, as companies can take measures to reduce or avoid them if possible. In this case, a clear and efficient product return policy, along with a possible punishment for the distributor, could help reduce the probability of receiving a product in poor condition. Some people may also think that a delay in product arrival is a punishment, but in reality, it is not. The delay of the reinforcer affects the value of the reinforcer, not its reinforcing nature. This effect occurs when the time between the behavior and the administration of the reinforcer is greater than expected. This process is called delay discounting, which indicates the decay of the value of the

Table 3.1 *Reinforcer array of the molar behavior "purchasing a book for university classes through Amazon"*

	MOLAR BEHAVIOR				
S^D	R				S^{C+di}
Document containing the subject's bibliography	Purchase book on Amazon				Book

	ARRAY OF REINFORCERS				
S^C Description	Tipo Consequence	S^C Contingency	S^C Contiguity	S^C Mediation	Term
1. Not using the car	Reinforcer	Negative	Immediate	Indirect	noS^{C-ii}
2. Not interacting with anyone	Reinforcer	Negative	Delayed	Indirect	noS^{C-ii}
3. Receiving the book	Reinforcer	Positive	Delayed	Indirect	S^{C+di}
4. Finding the book on the platform	Reinforcer	Positive	Immediate	Direct	S^{C+id}
5. Book in poor condition	Punishment	Positive	Delayed	Indirect	S^{C-di}

reinforcer as a function of the waiting time to obtain it.[6] This effect is clearly observed in food delivery. If we order food through a delivery app, and it arrives cold because it took longer than expected, the food still satisfies hunger, but its reinforcing value has decreased, both because of the taste and the increased expected wait time. The decay in the value of the reinforcer over the time elapsed between behavior and reinforcer follows a mathematical function known as hyperbolic decay function (Mazur 1997). In this case, the user may try other competing apps or consider other alternative behaviors before using a service that administered a reinforcer with lower value than expected again. Amazon, for instance, has been very aware of the risk that product delays could pose to its business, which led to the creation of Amazon Prime to reduce the effect of delay discounting, and the impact it could have on sales. Therefore, in addition to punishments, it is also interesting to observe how reinforcers are administered to find potential areas of opportunity for business improvement.

Let us consider another example, this time with the music streaming app, Spotify. In this example, the young David has to go to work, and for this, he steps out onto the street with his headphones listening to music as he makes his way to his destination. In this case, the molar behavior would be listening to music, and the discriminative stimulus would be the exit door of his house. The consequent stimulus would be the music, but just as with buying a book from Amazon, many other reinforcers are affecting the behavior of listening to music on Spotify while moving to his destination. What might these be? (see Table 3.2).

Unlike the case with Amazon's service, here the majority of reinforcers are direct, that is, obtained directly from the technological tool where Spotify is installed: listening to music (immediate direct positive

---

[6] The delay of a reinforcer affects the perception of its value. These effects are known as "delay discounting," defined as the cognitive process that allows people to compare the value of obtaining a reinforcer immediately or after a delay (Chung & Herrnstein 1967). Various theoretical approaches explain that the reduction in value of a delayed reinforcer is related to the time before it is delivered (Tesch & Sanfey 2008), or the subjective depreciation as a consequence of the delay (Baker *et al.* 2003), or the preference of individuals for small and immediate reinforcers over large and delayed ones (Mar & Robbins 2007), or the preference for any immediate reinforcer over a delayed one (Carroll *et al.* 2009). Although delayed reinforcers retain the capacity to increase the likelihood of the individual repeating the behavior, they are generally less effective in establishing new behaviors or modifying previous ones than immediate reinforcers (Lattal & Gleeson 1990; Sutphin *et al.* 1998). Moreover, caution is warranted if a delayed reinforcer is introduced in behaviors that were previously reinforced immediately, as the subject's response rate may decrease (Lattal 2010). Nonetheless, it is not yet clear what variables determine the value that drives an individual to act, rather than contemplating the valuable object (Keramati & Gutkin 2014; Levy & Glimcher 2012; Pelletier *et al.* 2021).

Table 3.2 *Reinforcer array of the molar behavior "listening to music on Spotify with headphones during the commute to work"*

MOLAR BEHAVIOR		
S^D	S^D	S^D
The sound made by the door of one's house when closing	Listening to music on Spotify	Music

ARRAY OF REINFORCERS

S^C Description	S^C Description	S^C Description	S^C Description	S^C Description	Term
1. Listen to music	Reinforcer	Positive	Immediate	Direct	S^{C+id}
2. Discover new songs	Reinforcer	Positive	Immediate	Direct	S^{C+id}
3. Stop thinking about problems	Reinforcer	Negative	Immediate	Indirect	noS^{C+ii}
5. Not hearing street noises (increased likelihood of accidents)	Punishment	Negative	Immediate	Indirect	noS^{C+ii}
6. Hear the street noise	Punishment	Positive	Immediate	Indirect	S^{C-ii}

reinforcer), discovering new songs (immediate direct positive reinforcer); as well as other indirect ones, such as ceasing to think about problems (immediate indirect negative reinforcer). Punishments are also identified, such as not hearing street noise (immediate indirect negative punishment) and hearing the noises of the street (immediate indirect positive punishment).[7] Once again, these punishments were a business opportunity, but for whom? This time it was for the companies controlling the technological tools, Apple, Samsung, Sony . . . These companies developed an algorithm or system for their headphones that could filter out exterior noise both to cancel it and to amplify it. Punishments, once more, became a business opportunity.

3.5 How to Build a Digital Operant Box

In the seminal work titled "Adaptative Behavior and Learning" by the esteemed American psychologist and global authority on learning, John Eric Rayner Staddon, a straightforward framework is presented on how an organism interacts with its environment (see Figure 3.5).

The organism emits a behavior within a specific context, which can be observed and measured quantitatively. Following the behavior's emission, the organism is capable of obtaining a reinforcer, which may appear after a single behavior, the repetition of the same behavior multiple times, after a certain period without exhibiting the behavior, among other modalities. In controlled settings such as a Skinner Box or in a DOB, the administration of reinforcers can be categorized using various "reinforcement schedules." Depending on the reinforcement schedule utilized, the response rate or its resistance to extinction can vary. Furthermore, it is also possible to design whether once the reinforcement schedule's requirements are met, the reinforcer is administered immediately after the response or at different delay intervals (Staddon 2003). Staddon's model can also be employed to explain DOPs (see Figure 3.6). The organism would be a user behaving in a digital space through a series of actions, which can be parameterized and measured. These digital actions, broken down into parameters, are recorded by a set of algorithms specifically established in the "back end," which are programmed to associate a digital consequence with a specific digital action. For instance, if I click on the Shein digital store on a blouse

---

[7] Here it may seem contradictory, but at different times during the journey, the sound of the street noise can be a punishment ("it prevents me from hearing my music clearly") or not hearing it can be ("I can hurt myself by not perceiving possible dangers").

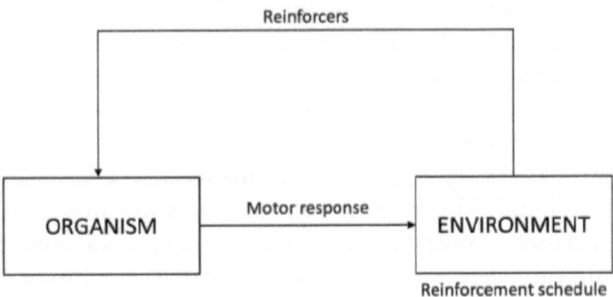

Figure 3.5 Feedback relationship in an operant conditioning experiment

In the context of operant conditioning, the feedback relationship pertains to how the organism is affected by the reinforcer produced by its motor response. The reinforcer, in turn, is governed by a reinforcement schedule, which, in operant experimental designs, is programmed by the experimenter. This reinforcer is delivered when the organism emits a specific motor response, with the environment exerting a feedback function to control behavior. The timing and consistency of the reinforcer, as determined by the reinforcement schedule, play a critical role in either strengthening or weakening the response over time. Together, these elements shape the relationship between the response and the reinforcer, which subsequently influences the likelihood of the response being repeated in the future (Staddon 2003).

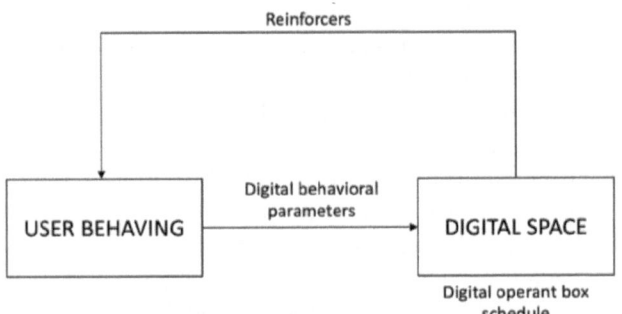

Figure 3.6 Basic design of a digital operant box

Users engage in behaviors within a digital environment, which are controlled by antecedent and consequent digital elements to the behavior (Skinner referred to any elements of operant conditioning as "operants"). The technological service must have been developed to quantify digital behavior across various parameters so that consequences to digital actions are activated according to a digital reinforcement schedule. Upon detecting the parameter that triggers a specific rule of the digital reinforcement schedule, the digital tool will administer a sensory element that, due to contingency and contiguity to the digital behavior, allows the user to perceive that the outcome is a consequence of their behavior. This functional relationship between behavior and consequence will consolidate this response for the next time the user seeks to obtain the same consequences in a similar context.

(digital action/behavior), the characteristics of the garment appear (immediate direct positive molecular reinforcer). The consequence of digital actions, as with actions in natural environments, has short- and long-term consequences.[8]

In digital contexts, it is typical for each user action to be followed by a corresponding consequence. This reinforcement schedule is known as a "continuous reinforcement schedule": one behavior equals one consequence. However, behavioral sciences have long utilized various programs that do not adhere to a 1:1 logic, instead requiring the organism to undergo diverse conditions to receive reinforcement. For instance, not every job application sent is rewarded with an interview or the job itself, nor does checking email or social media always yield new messages or notifications. A digital exemplar of how different methods of obtaining a reinforcer can be designed is through interactions within Facebook groups. Facebook has devised a type of reinforcer in the form of badges for users who interact within Facebook groups. However, these are not awarded merely for user interactions, they also require an additional criterion: user interactions must be accompanied by responses from others to their posts. In essence, Facebook monitors the reactions your post elicits within the rest of the user community in the group. By doing so, it ensures that it is reinforcing those users who are valuable in generating group interactions, as opposed to users who post content that fails to significantly resonate within the group – significance here being understood as prompting further user interactions. Facebook employs these individuals as "attractors" who enhance the participation of the community at large. To this end, it designed badges, yet the rule for obtaining them remains concealed. It is only known that they are awarded for active participation in the group and, additionally, if such involvement prompts further interactions from other group members. This type of schedule is known as a "variable reinforcement schedule," with the requirement to generate community participation, which is measured through interactions such as likes, comments, and so forth. The rule is clear: the more interactions with the platform, the higher the likelihood of engagement (Lindström *et al.* 2021). Therefore, the behavior of posting content to the group is reinforced, as well as the impact that the content has within the group (see Figure 3.7).

---

[8] The acquisition of a reinforcer and its feedback function can change an individual's internal state (Hull 1943). This change can be emotional and have short-term effects, as it may induce an emotional shift that will affect their subsequent interactions. Alternatively, it can be cognitive and have long-term effects, given that the relationship between S^D-R-SC is stored in memory to be utilized later if necessary (Gershman & Daw 2017; Rolls 2004; Rostami Kandroodi *et al.* 2021).

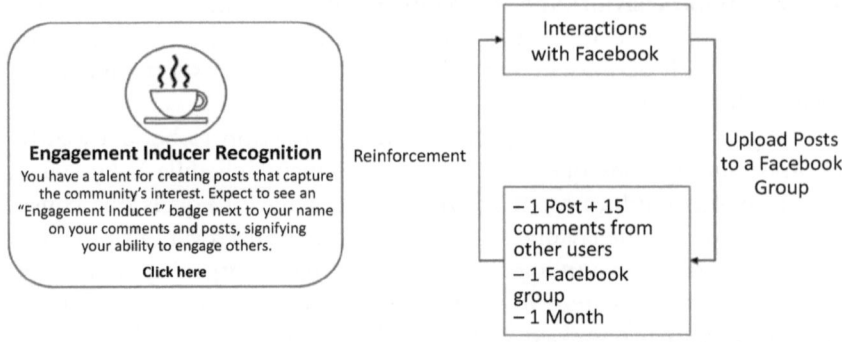

Figure 3.7 Facebook reinforcer in the form of a badge
Facebook is programmed to detect posts by users in groups that are capable of generating a high level of interactions. Users who manage to elicit interactions from other group members around their posts are identified by the system as valuable users. In order to maintain and encourage these valuable users to continue their activity, Facebook delivers this reinforcer in the form of a badge. The system's rule for managing these reinforcers may be following a conjunctive reinforcement schedule with a requirement of interactions per post. That is, the user must post at least: (1) 10 posts a month with 15 interactions, (2) in a Facebook group, (3) with a validity period of 1 month to receive the reinforcer in the form of a badge (the figure provided in the example are a representation of the actual ones).

It is also crucial to note that this badge as a reinforcer serves other functions beneficial to the Facebook platform. The badge is displayed for a month in the comments made by the user who obtained it. In this way, other users can learn through vicarious learning how to obtain this reinforcer if they emulate the behavior of those with badges. They may even delve into their profiles to examine the frequency and nature of the posts. Content creators indeed correlate the interactions with their material uploaded to the networks to the content therein. It is a meticulous study, where these influencers have at their disposal an abundance of tools providing numerical data on the interactions with the various contents they have shared. In this process of study, where interactions are linked with the type of content, the phenomenon of shaping emerges. This operant learning process indicates that the progressive and successive obtaining of reinforcers can gradually approximate a desired final behavior. That is, likes signal the type of content audiences desire, so the influencer, wishing for more likes (dopamine release) to reach a wider audience and to feel happy with their work, plans the next content similar to those that garnered more interactions. Consequently, over time, interactions will shape the content of their publications until they find their market niche. This is paradoxical because it is the market or audience that creates influencers through their interactions, and the influencer, in turn, alters the audience with the content of their posts (Guilbeault & Centola 2020). However, these shaping interactions can have pernicious effects, such as skewing the influencer toward increasingly extreme ideas with consequences such as the expression of moral indignation or "moral outrage expressions" (Brady *et al.* 2021).

3.6 Digital Reinforcement Schedules

Specifically, and as previously mentioned, the way in which the response-consequence relationship (specifically in its contingency relation) is structured is known as a reinforcement schedule (Skinner 1938). Reinforcement schedules are a set of rules that determine which forms of responses will be reinforced, with the frequency of the response and the time elapsed between responses being the main variables. In the design of digital behaviors, these will be referred to as "Digital Reinforcement Schedules" (DRS), which will operate under the same principles as reinforcement schedules, except that the reinforcers will be administered by a digital service in a digital environment. DRSs are defined by a set of rules that determine the administration of reinforcers following the emission of certain specific user behaviors in a digital environment. These schedules are designed prior to the development of the service by a digital behavior designer, in such a way that they digitally simulate the contingencies that occur in natural contexts. There are two main modalities in which a reinforcer can be administered: continuous and intermittent. Continuous reinforcement schedules are those rule systems in which a reinforcer is administered after every response from the organism; whereas intermittent reinforcement schedules are those systems in which not every response is followed by a reinforcer. Within the intermittent modality of administering a reinforcer, there are various schedules based on the number of responses from the organism and the time elapsed since the last time a behavior was reinforced (see Figure 3.8).

The intermittent reinforcement schedule is divided into two types, ratio and interval. Ratio schedules are those that depend on the number of responses emitted by the organism, while interval schedules are those that depend on the time elapsed until the reinforcer becomes available again from the administration of the last reinforcer. Both ratio and interval

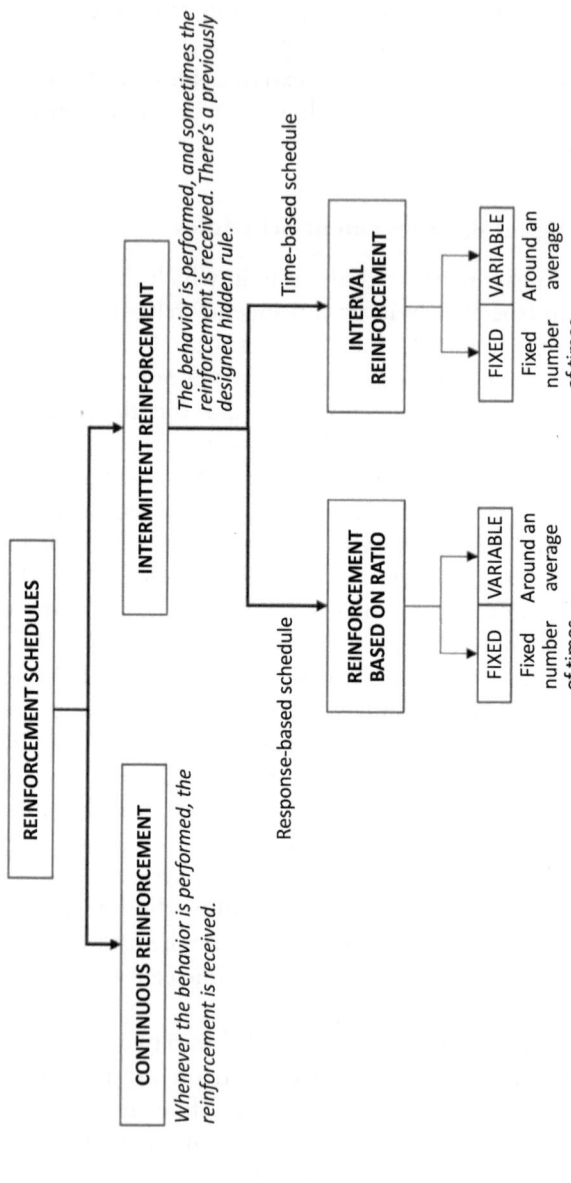

Figure 3.8 Outline of primary reinforcement schedules

Reinforcement schedules are divided into two broad categories: continuous reinforcement schedules (whenever I perform the behavior, I obtain the reinforcer) and intermittent reinforcement schedules (I perform the behavior, and sometimes I obtain the reinforcer). There is an underlying rule designed a priori in intermittent schedules. In turn, the intermittent reinforcement schedule is further divided into two types based on whether it is predicated on the number of responses emitted by the organism (ratio schedules) or the time elapsed between the administration of reinforcers (interval schedules). Subsequently, each of these can be based on a fixed number of responses/time, or a variable number of responses/time that revolves around an average.

Table 3.3 *Description of continuous and intermittent reinforcement schedules*

Reinforcement Schedules		
Type	Description	Effect on Behavior
Continuous Reinforcement Schedule.	Each response is reinforced.	Responses occur at a constant and moderate rate.
Fixed-Ratio Reinforcement Schedule.	Reinforcement is administered after the emission of a fixed number of responses.	The responses are maintained at a high and constant rate, accompanied by a brief pause after obtaining the reinforcer.
Variable-Ratio Reinforcement Schedule.	Reinforcement is administered after the emission of a variable number of responses centered around an average.	The responses occur at a fairly constant rate, without being able to predict a pause after obtaining the reinforcer.
Fixed-Interval Reinforcement Schedule.	Reinforcement is administered with the emission of the first response after a fixed time has elapsed since the last reinforcement administration.	The responses appear toward the end of the interval, and then cease until the interval time is approaching its end again.
Variable-Interval Reinforcement Schedule.	Reinforcement is administered with the emission of the first response after a variable time centered around an average has elapsed since the last reinforcement administration.	The responses maintain a constant and stable rate without regular pauses.

schedules can be of a fixed or variable type. Below are the characteristics of each type of reinforcement schedule (see Table 3.3).

Digital service designers can design the number of responses that the user must emit or the time that must elapse until the user can again obtain the next reinforcer. Below we show some examples of the administration of real and digital reinforcers through the different reinforcement schedules (see Table 3.4).

Table 3.4 *Natural and digital examples of continuous and intermittent reinforcement schedules (some data regarding the number of behaviors or time required for the reinforcer to appear are fictional to facilitate understanding of the example)*

Reinforcement Schedules	Examples of reinforcement in natural environments	Examples of reinforcement in digital environments
Continuous Reinforcement Schedule.	A dog receives a treat each time it detects explosives.	Every time I order food through Uber Eats, it is delivered to my address.
Fixed-Ratio Reinforcement Schedule.	A salesperson at a publishing company receives a commission every five sales.	Every 3 times I watch a video of an influencer on TikTok, the frequency of their videos appearing on my feed increases.
Variable-Ratio Reinforcement Schedule.	Casino slot machines are programmed to give a jackpot every certain number of spins that varies around an average.	Uber offers discount coupons after completing several rides, sometimes 4, other times 5, 6, or 2.
Fixed-Interval Reinforcement Schedule.	A weekly paycheck.	In Nintendo's Mario Kart Tour, every 24 hours, a new challenge is given to complete.
Variable-Interval Reinforcement Schedule.	A teacher gives surprise exams, at least one every two weeks.	Every 1, 2, or 3 hours, the opportunity to participate in a special match with unique prizes appears in Fortnite.

An application that has adeptly utilized DRS is Candy Crush, a video game developed by the British company King Entertainment for mobile devices. It is a puzzle and strategy game, whose goal is to match three or more candies of the same type to make them disappear and score points (it is recommended to play or view the game on the internet before continuing to read). Part of the reason this game is so addictive (255 million users in 2021) (Business of Apps 2023) is due to its delivery of multiple reinforcers at different stages of the game. For example, a continuous DRS is when each time you match 4 identical candies in a row, you receive a "booster" that increases your destructive power within the game. Fixed-ratio DRSs are also common, such as every time you surpass fifteen levels (behaviors), you gain a new world to play in (reinforcer); or every time you win three consecutive games without losing a life, Candy Crush awards a crown (reinforcer) that appears above your game avatar. A variable-ratio DRS occurs when Candy Crush gifts a life (there are limited lives in the game that, when exhausted, prevent further play until a certain amount of time has passed) every time a Facebook friend accepts your invitation to join the game. Here, a variable number of "send invitation" behaviors must be performed for a Facebook friend to accept the invitation to Candy Crush, hence granting you an additional life in the game (reinforcer). Another example is the number of moves required to pass a movement level (this is how such levels are named). That is, sometimes you need twenty moves, other times thirty-five, and yet other times fifty to advance, similar to a rat needing to press a lever twenty, thirty-five, or fifty times to receive food. There are also many fixed-interval reinforcement schedules, such as receiving a free "booster" daily in a game known as the wheel. Finally, the variable-interval DRS is more difficult to find in the game, as many of the game rules are kept hidden and secret. A possible variable-interval DRS could be when the game itself offers help after a time playing at a level. Sometimes this assistance appears at 120 seconds, other times at 100, and another at 140 seconds, possibly depending on the average time it took other users to give up trying and abandon the game completely.

Another example to which digital behavior designers must also pay attention is how the digital service affects real life. In this sense, the very activity of playing Candy Crush could act as a reinforcing stimulus for many people as well (Premack Principle). Many might take advantage of the "free" spaces in their schedules to play Candy Crush, such as during public transport commutes, waiting rooms, or even breaks at their regular jobs. Retail workers might play Candy Crush during lulls when no customers are visiting the store. This is an example of a natural variable-interval schedule, as the flow of customers (and thus the breaks for playing) would fluctuate according to certain external variables, so access to the reinforcer would be determined by the moments when there are no customers in the store. Moreover, this schedule is natural because it has not been specifically designed by anyone, although the average time between reinforcers could be calculated retrospectively if we measured the times between breaks over a certain period minutes (months, years, seasons . . .). Another interesting phenomenon to mention in Candy Crush is a particular type of discriminative stimulus that appears when you are close to passing a level and lose a life. At that moment, Candy Crush shows the necessary moves to finish the level. The display of these moves can be interpreted as a discriminative stimulus that anticipates that if a life is purchased, the user could pass the level (reinforcer). On the other hand, these discriminative stimuli can also be interpreted as a near-miss, which would have reinforcing properties that exert a powerful effect on the user's motivation to repeat the behavior of playing again. Near-misses cause greater physiological and subjective arousal than when it is not indicated how close one was to obtaining the reinforcer (in this case, the next level of the game) (Barton *et al.* 2017). This type of signaling causes great frustration that triggers the need to continue playing. That is, indicating how close one was to accessing the anticipated reinforcer causes users to play for longer than expected (Larche *et al.* 2017). Therefore, understanding the functioning of reinforcement schedules and how they operate on behavior will be one of the basic tools for the designer of digital behaviors.

3.7   Digital-Specific Adaptations of Reinforcement Schedules

Lastly, it is also appropriate to make certain clarifications concerning the different DRS that can be designed, which may cause conflicts when trying to categorize them as one type or another. Many examples found on social networks suggest, for instance, that when a child has been studying for an hour and is then granted a fifteen-minute break, this constitutes a Fixed Interval Schedule. In the phrase "when the child has been studying for an hour," it refers to an actual amount of study time, whereas Fixed Interval Schedules do not consider the duration of an organism's activity but rather that the activity occurs after a specific period. It might seem similar, but the nuance is significantly important. Pragmatically speaking, there is a vast difference between studying for sixty minutes versus five minutes, hence a Fixed Interval Schedule fails if the designer's objective is indeed for the child to study for sixty minutes. If it ultimately gets classified as a Fixed Interval Schedule of sixty minutes, where the behavior is studying and the reinforcer is a fifteen-minute break, during the inter-reinforcer interval, the student is not obliged to study. Simply beginning to study after the sixty minutes would suffice to receive the reinforcer. Within the inter-reinforcer interval, various behaviors may manifest that would be considered induced behavior and are categorized as adjunctive behaviors, but they appear randomly and are not a requisite for the administration of the reinforcer. In the example of the student, the break would actually be administered irrespective of whether the child has been effectively studying for sixty minutes. If the true intention is to reward actual study time, this example should be regarded as a Continuous Reinforcement Schedule with a time requirement (Beer & Trumble 2014). If there are pauses before reaching the sixty minutes, it is the designer's decision whether to reset the timer to zero, or pause it until the individual resumes the target operant response of studying for sixty minutes. This type of requirement falls under the well-known Duration-Based Reinforcement Schedules, which may be Fixed (Fixed-Duration Schedules) or Variable (Variable-Duration Schedules) (Gulotta & Byrne 2015). Fixed-Duration Schedules are those where the actual time spent engaged in the behavior is constant, while Variable-Duration Schedules are those in which the actual time varies around an average. There are even Progressive-Duration Reinforcement Schedules, in which the actual time requirement to perform a behavior increases as the person consistently achieves the reinforcers. An instance of these schedules can be seen in games where the playing time is the behavior enabling the acquisition of certain reinforcers. A Fixed-Duration Reinforcement Schedule could be when a new upgrade for an avatar is offered every three hours of effective gameplay. Conversely, a Variable-Duration Reinforcement Schedule might occur when an upgrade is provided sometimes after the user has played for two hours, other times after three or four hours, with the average being three hours. Finally, a Progressive-Duration Reinforcement Schedule is a Fixed-Duration Schedule that increases its duration limit as the player obtains reinforcers. The goal is to gradually "guide" the player toward creating a habit of extended gameplay duration.[9] Starting with rewarding short gaming times, fifteen minutes and then progressively to thirty minutes, one hour, two hours ...[10] On the other hand, other typical behaviors of digital service users that can be rewarded may be associated with different parameters than response rate, time between reinforcers, or behavior duration. For example, in the Uber app, rewards could be given for accumulated kilometers from trips made by a user. Here, the digital behavior designer must determine which behavior is associated with the creation of kilometers. In this case, it is the behavior of trips made through the app.

---

[9] Similarities have also been found between progressive reinforcement schedules based on response and duration, in the sense that increasing the response requirements after each reinforcement delivery can ultimately result in the cessation of the response (Bailey *et al.* 2015; Gulotta & Byrne 2015). Even if the duration requirement is removed, the time during which the organism will perform the behavior might also decrease (Byrne & Sarno 2019).

[10] For greater conceptual precision, it is important to distinguish between the progressive duration reinforcement schedules and the operant technique of shaping. In shaping, reinforcement is given for the progressive performance of a behavior, regardless of its temporal duration, unlike the progressive duration schedules, where reinforcement depends exclusively on the execution time of the specific behavior.

Thus, we are looking at a Variable Ratio Schedule with a kilometer requirement, as we do not know the number of trips each user will need to accumulate to obtain the reinforcer. This scenario is common in games, even with a progressive design. For instance, in massive multiplayer online role-playing games like Ultima Online, improving an avatar's skill becomes progressively more challenging to obtain. This difficulty translates operationally into the player needing to perform more behaviors related to the skill as they advance or reach a new level (reinforcer). If initially, only writing five scrolls is needed to level up the scribe skill (Fixed Reinforcement Schedule), the next level might require performing the scroll writing behavior ten times, and so on, until reaching the maximum level. Depending on the designers, this type of digital behavior for avatar skill enhancement can be classified as a Progressive Fixed Ratio Schedule, or even a Multiple Schedule where various Fixed Ratio Schedules are concatenated. Meanwhile, some professionals in the field, due to the significance of these schedules within their frameworks, may refer to them as Kilometer-Based Rewards Systems (Greaves & Fifer 2010) or even Point-Based Reinforcement Systems (Ahn *et al.* 2019). However, it is vital for the designer of digital behaviors to be aware that these molecular reinforcers (kilometers or points) are associated with the execution of a specific behavior, thus they can always be referred to as ratio schedules with some type of requirement.

Less common, yet equally significant, are conjunctive, multiple, and concurrent reinforcement schedules. Conjunctive reinforcement schedules are those in which the criterion for obtaining the reinforcer is determined by the simultaneous fulfillment of at least two previously mentioned reinforcement schedules. For instance, in a video game, to fully replenish your avatar's energy bar, you need to defeat ten enemies (fixed ratio) within a span of twenty seconds (fixed interval). In contrast, concurrent reinforcement schedules occur when the user can choose between two or more reinforcement schedules that are available simultaneously. These types of schedules can be easily found in the daily lives of users when they have limited time and must decide between using TikTok or Instagram during their leisure time. There are also multiple reinforcement schedules where different reinforcement schedules are presented either in parallel or sequentially, each with its distinct discriminative stimuli. For example, in adventure video games where the avatar goes through various challenges, one could design based on the implementation of a multiple reinforcement schedule, where each stage or level of the game corresponds to a different reinforcement schedule. Lastly, another reinforcement schedule known as "fixed time" is to be explained. Unlike the interval schedule, where the reinforcer is administered when the user performs the first behavior after a time interval, in the fixed-time schedules, the reinforcer is administered after a predetermined time, regardless of whether the user performs the target behavior or not. In the game Candy Crush, users receive a life every thirty minutes, up to a maximum of five. The delivery of this life is independent of what the user does, even if they have been playing or not. Yet, in the game of Candy Crush, the user does not necessarily have to wait thirty minutes to obtain another life; instead, they can purchase it in the online store. Therefore, the Candy Crush user is presented with a concurrent reinforcement schedule, where they can buy a life (continuous reinforcement schedule) versus waiting for thirty minutes (fixed-time schedule). In economic terms, reinforcement schedules can also encourage behaviors that are profitable for the company, without blocking the use of the service (see Figure 3.9).

3.8 The Potential Adverse Effects of Reinforcers

To date, the significance of designing effective reinforcers and discriminative stimuli has been underscored to enhance the likelihood of a user engaging in the target behavior within a digital business (such as viewing specific content, making a purchase, or spending time with the product). Indeed, it appears that the mere identification and selection of a reinforcer ensures success in the establishment of molar behavior. However, this is not always the case, and the digital behavior designer must project or simulate potential parallel reinforcers that could emerge alongside the implementation of the principal reinforcer. These parallel reinforcers might overshadow the value of the primary reinforcer and lead to behaviors that are undesirable for the business. An instructive case of poor reinforcer design was demonstrated by the incident with Domino's Pizza and its "thirty minutes or it's free" promotion. This promotion offered the pizza (reinforcer) for free if it was delivered in more than thirty minutes. That is, the campaign's aim was to decrease waiting time (delay discounting effect), thus enhancing the value of Domino's pizza compared to its competitors. Hence, the greater the value of the reinforcer, the higher the probability of choosing Domino's over other competitors such as Pizza Hut. Aside from customer cunning to get free pizzas, the obligation to meet delivery times prompted Domino's delivery personnel to increase their driving speed, which was an unintended consequence that led to an increase in traffic accidents involving their delivery drivers. Furthermore, a $79 million

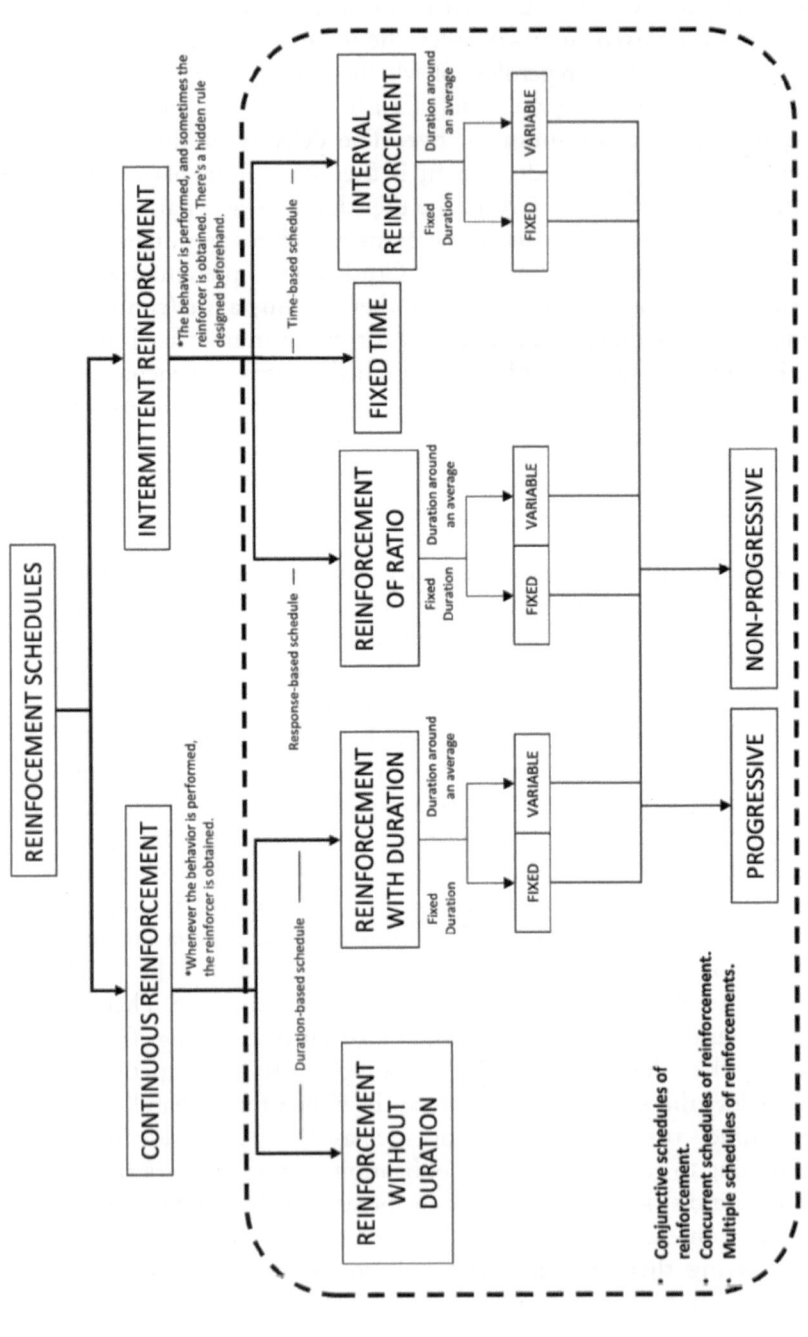

Figure 3.9 Expanded schema of reinforcement schedules

Traditionally, reinforcement schedules, as depicted in Figure 3.8, have formed the core understanding of behavior shaping mechanisms. This figure presents an expansion of the aforementioned schema by introducing other equally important schedules for the design of digital behaviors: conjunctive, concurrent, multiple, response duration-based, fixed-time, and progressive reinforcement schedules.

lawsuit ensued, forcing Domino's to withdraw the promotion (Janofsky 1993). Another interesting example of poor design occurred at a school that began to financially penalize parents for late pickups of their children after class had ended. The school's objective was for parents to collect their children earlier to reallocate school resources to other tasks or even to close earlier to reduce costs. Over time, parents began to regard this penalty as an additional child care service provided by the school, rather than a punishment. What was the outcome? Parents started to pick up their children even later than before. Lastly, there are examples in the digital realm as well. For instance, the anonymous messaging application Yik Yak, which initially saw great success, had to shut down after four years because the medium-term consequences of its primary reinforcer, message anonymity, were not measured. As time progressed, many messages on the platform shifted to a more aggressive tone, which spawned racist behaviors and cyberbullying (Safronova 2017). Another misstep in reinforcer design was with the video game Punch Quest, which continues to operate but was a financial disaster. Despite having over 630,000 users, it only generated $10,000 in profit. Why? They offered the majority of valuable reinforcers for free, resulting in users seeing no reason to pay for other options. Perhaps this decision to offer many reinforcers at no cost was coherent initially with the goal of rapidly acquiring many users, but it also led to significant losses that jeopardized the company.

CHAPTER 4

Anticipating Anticipation

4.1 The Concept of the Mind

The concept of the mental is among the most contentious issues within psychology. The notion of the mind met significant resistance in the first half of the twentieth century due to the numerous experiments on the nervous system and animal behavior, which yielded results that did not require invoking intentional internal states, such as the mind, to explain and predict behavior. This approach to the study of psychology evolved with the emergence of new evidence, which began to point to the existence of a mediating element between the environment and behavior – mental or internal representations (Holland & Rescorla 1975; Rescorla 1973, 1974; Rescorla & Cunningham 1978). Coupled with this body of research that emerged from the second half of the twentieth century, along with other studies more related to the activity of the nervous system, an alternative psychological perspective to behaviorism was emerging: cognitivism. Cognitivism is currently the prevailing theory on human psychology, although paradoxically, no one knows exactly what it is (Bayne *et al.* 2019). There is a consensus in the scientific community that cognition refers to computational processes or mechanisms of the brain (Marr 1982) that operate on the contents of consciousness (Liu *et al.* 2021), so that it is possible to simulate plausible potential futures (expectations) that will ultimately control our final behavior (Blom *et al.* 2020). In the design of digital behaviors, the expectations that users bring to the use of the digital service also have to be designed based on cognitive models provided by psychology and neuroscience. The mentioned expectations are based on the anticipations that the user has about the use of the service. These anticipations, in turn, derive from the user's learning history, which consists of previous experiences acquired through past interactions with the service, as well as the observation of other people who have used it. A mismatch between the expectations of using the digital service and the reality of its use can generate unpleasant emotions that negatively affect future interactions. To understand how the anticipations of using a technological service affect the user's digital behavior, they should not be interpreted solely as a consequence of the environment but also of the mental aspects.

4.2 The Cause of Digital Behavior

To anticipate people's behavior, it is necessary to understand the causes that originate it. As explained in previous chapters, according to behaviorism, the cause of behavior lies in the environment, denying any participation of the individual's internal states. According to these theoretical approaches, the environmental stimulus would be directly associated with patterns of muscular sequences, so that the mere presence of a stimulus could trigger the individual's behavior, without any kind of mediation. One of the authors who possibly best reflects how futile it would be to search for the internal causes of behavior is the Russian physiologist Ivan Petrovich Pavlov, whom we will discuss later:

> Since we used the studies of the lowly organized representatives of the animal kingdom as an example, and, naturally, wanted to remain physiologists instead of becoming psychologists, we decided to take an entirely objective point of view also towards the psychical phenomena in our experiments with animals. Above all, we tried to discipline sternly our way of thinking and our words and ignored completely the mental state of the animal; we restricted our work to careful observation and exact formulation of the influence exerted by distant objects on the secretion of the salivary glands. The results were according to our expectations: the observable relations between external phenomena and variations in the activity of glands could be systematically analysed; they appeared to be determined by laws, because they could be reproduced at will. We were pleased to find that our experiments proved to be right and fruitful. I shall give some examples here illustrating the results obtained with the new method in our field of interest. (Iván P. Pávlov – Nobel Prize Speech (1904))

Pavlov did not deny the existence of the mind or the subjective states of the individual; rather, he refuted the necessity of understanding them to predict and anticipate the behavior of organisms.[1]

---

[1] To predict and anticipate denotes the ability of a system to possess foreknowledge of an event's cause-and-effect relationship. However, in this text, anticipate is used as a distinctly human characteristic due to the cognitive aspects that are involved.

I do not deny psychology as a body of knowledge concerning the internal world of man. Even less am I inclined to negate anything which relates to the innermost and deepest strivings of the human spirit. (Iván P. Pávlov – Lectures on Conditional Reflex (1928))

The high precision with which Pavlov and his team were able to predict the responses of animals was surely related to the rise of behaviorism in the United States (Windholz 1983), and hence, the significant position this school of thought occupies within psychology. Indeed, psychology is considered a science and not merely a philosophy thanks to the findings of this group of scientists who were able to mathematize behavior, make predictions, and establish a categorical closure around the study of organism behavior.[2] Behaviorist principles can be regarded as the foundation upon which behavior is constructed, although discoveries made over the last century have led to explanations of behavior that transcend the environmental to reach the mental.[3] To provide clarity to this chapter, certain basic terminology that may be confusing for professionals from various branches of psychology will be defined as follows:

• Neuronal Representation: A neuronal representation is the configuration of an environmental or internal stimulus within a neuronal space created from the encoding of this object (Piccinini 2022). This neuronal space forms a nervous structure composed of different sensory, motor, emotional, and mnemonic dimensions that can constitute from a broad concept to an environmental element such as a physical space, a stimulus, or a word. This neuronal structure is dynamic in nature, that is, it changes as the organism interacts and acquires different environmental experiences (Binder *et al.* 2009; Constantinescu *et al.* 2016).

• Mental Representation: A mental representation refers to a neuronal representation that is activated in working memory,[4] which the subject can manipulate and of which they are subjectively aware.

---

[2] Gustavo Bueno Martínez's theory of categorical closure, proposed by the Spanish philosopher, advocates from a scientific pluralism perspective that each scientific epistemological field should have its own tools to study reality which are not related to the researcher's subjectivity.

[3] The adjective "mental" refers to the subjective experience that emerges from the integrative activity of different neuronal circuits which facilitate the cognitive operability of information encoded in their own circuits with the ability to impact behavior.

[4] The neuronal representation that is activated in the working memory, and which the individual is capable of accessing metacognitively for its manipulation, will be referred to in this chapter as "cognitive unit."

- Memory: In general terms, memory is the cognitive anchoring of an experience in a neuronal representation. This memory can be explained in its procedural dimension (encoding, consolidation, and retrieval), its temporal dimension (working memory, intermediate memory, or long-term memory), or qualitative dimension (declarative or nondeclarative memory) (Squire 2004).
- Cognitive Units: Cognitive units are a term coined by Anderson in the 1980s, referring to a segment of long-term memory activated in working memory. The theory of cognitive units suggests that information is stored in long-term memory in the form of coherent and related units, which are activated in working memory when required for specific cognitive processes. By activating a cognitive unit, one can access related information and use it for specific cognitive processes such as problem-solving or decision-making. These units can vary in size, from simple elements like letters or numbers to more complex ones like words, phrases, images, odors, or emotions among others. In the design of digital behaviors, these elements that make up a cognitive unit are known as nodes (Anderson 1980). Each of these nodes is a neural representation in itself. The connections between the nodes that form the cognitive units will be referred to as "associative traces," which define the associative relationship and intensity among several nodes of the cognitive units (Rescorla & Wagner 1972). Defining this strength is crucial since as the associative trace increases, the likelihood of activating one of the elements leading to the complete or partial activation of the associated elements through signal propagation also increases (Powers *et al.* 2017). This propagation may extend to nodes within the same cognitive unit or to nodes of other cognitive units (Anderson 1983a; Foster *et al.* 2017a, 2017b). The concept of cognitive units may appear similar to other concepts like gnostic units or chunks.

In the design of digital behaviors, the brain is conceived as a biological system that allows organisms to anticipate the future by detecting and storing patterns and environmental regularities. Anticipation processes are the anchorage of the mind to the material world. As will be discussed later, this anticipatory system is a dynamic system containing mental models of how it and the world function, which are constantly updated due to the influx of new sensory information into the system. The animal brain, and therefore the human brain, is designed to find patterns that regularly signal the appearance of an element of value to the individual and, in some way, to act accordingly.

The ability to anticipate and prepare for what is going to happen allows the organism to dominate new environments. For example, smoke may signal an aversive stimulus if it comes from a residential complex of houses; or an appetitive stimulus if the person is lost in the forest since it signals human activity. The greater the capacity to predict and anticipate, the greater adaptive power the organism will have. For both aversive and appetitive stimuli, the brain tends to prepare the human both physically and psychologically through a process known as allostasis (Sennesh et al. 2022).[5] But here arises a key question: How does an organism anticipate the future? When we experience a situation, our brain creates a "small clip" of the functional relationship between all environmental elements before, during, and after our behavior. If this "clip" contains new information, it could be considered a new experience, which is structured into an associative map of all the environmental functional relationships (Knudsen & Wallis 2021b; Shikano et al. 2021). This new associative map accommodates within already existing neuronal representations, altering its associative configuration (number and strength of associations) of the various elements/nodes that compose it, modifying the neuronal representation itself. This process will repeat throughout our lives as we experience new situations with new information (Dudai et al. 2015).

The appearance of environmental elements with which there has already been some type of experience can mentally activate these associative maps, allowing for the simulation of different futures that enable us to plan our behavior (Anderson 1983b; Vogel & Wagner 2018). Let's imagine the following case. A teenager named Stuart suffers bullying at school from an older student named Lucas during the break between classes. During the different bullying experiences, Stuart's brain has constructed associative maps of all the environmental elements that were present before, during, and after the bullying. All these elements associated with bullying are used as cognitive units in Stuart's working memory in such a way that they help him anticipate if he will suffer bullying, and therefore, propose avoidance behaviors. For example, the class-ending bell is one of the elements associated with the neural

---

[5] Allostasis is a cerebral process that enables the maintenance of the body's physiological stability while producing functional changes that allow us to actively adjust to predictable or unpredictable events. In this regard, the brain must constantly coordinate the various physiological systems of the organism. To carry out this task, the brain requires processing and integrating information from the internal and external world through its own models of the world. Psychology suggests that these internal world models are created from memory, thoughts, perceptual inferences, non-conscious inferences, embodied simulation, concepts and categories, controlled hallucinations, and predictions. The brain needs to construct a dynamic simulation of the organism in the world to be able to direct it toward the desired goal. Therefore, allostasis control is a process that allows adjusting the organism to the physiological needs of its own anticipations (Sennesh et al. 2022).

representation of Lucas, the bully, so Stuart anticipates that there is a risk of suffering bullying. The bullying has not yet occurred in reality, but in Stuart's mind, a cognitive unit that associates [bell-break-hallway-Lucas-bullying] is already activated. From this point, two processes occur in Stuart, one automatic and fast, and the other reflective and slow. As for the automatic and fast process, Stuart's brain prepares his body to act in case of threat by inducing an emotional activation response, which can be felt as anxiety (Barlow 2000). This response, in a way, prepares Stuart for action and will play a very important role in the reflective and slow process (Rescorla & Solomon 1967). The cognitive unit [bell-break-hallway-Lucas-bullying] is represented in Stuart's working memory, in such a way that it allows him to plan a scenario that manages to avoid the anticipated aversive stimuli (Lucas's bullying), and also, to eliminate the emotional state of anxiety automatically induced upon hearing the bell (Liddell 1949). In this case, Stuart may decide to stay in the classroom to avoid going out into the hallway and encountering Lucas. If, in the end, Stuart decides to stay in class, and that operant behavior avoids the aversive stimuli of bullying, this functional relationship of [bell (S^D) → staying in class (R) → no bullying (noS^{C-})] will accommodate as a new experience in the neuronal representation of bullying as a possible solution (avoidance conditioning) (Constantinescu *et al.* 2016) (see Figure 4.1).

The anticipation of an aversive stimulus enabled Stuart to generate a new behavior through which he managed to avoid harassment and, thereby, reduce the anxiety provoked by hearing the bell at the end of class. In the future, when the bell rings, Stuart will have the option to stay in class to evade harassment. The complexity of this example has been simplified, as there are numerous other associations that may interfere with Stuart's behavior during class. For instance, the need for the teacher's presence in the classroom to prevent harassment, since Lucas looks for him when he is not found in the hallway (see Figure 4.2).

Moreover, Stuart may have associated Lucas's motorcycle with the harassment within the same cognitive unit. That is, if Stuart does not see Lucas's motorcycle in the parking lot upon arrival at the institute, he anticipates that Lucas will be absent that day, hence there will be no harassment (negative reinforcement; S^{C-}), and he will be able to act freely (positive reinforcement; S^{C+}) (see Figure 4.3).

Another possibility is that when he steps into the hallway after class ends, he often engages in conversation with Judith, a person toward whom he feels a great attraction (positive reinforcer; S^{C+}). The value of this reinforcer far exceeds the value of avoiding Lucas, so he may decide to leave class, despite the multiple stimuli indicating the risk of harassment (see Figure 4.4).

A)

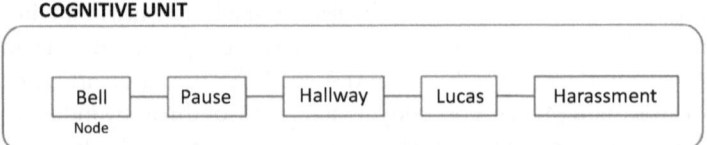

B)

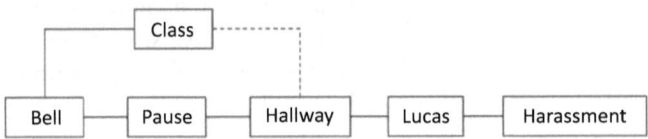

Figure 4.1 Representation of the cognitive unit of bullying 1

The excitatory connections indicate a positive contingent relationship (the likelihood that the contingent stimulus will appear); whereas the inhibitory connections indicate a negative contingent relationship (the likelihood that the contingent stimulus will not appear). **A)** Stuart is manipulating the cognitive unit in working memory that leads to Lucas being bullied. The contingent relationship presented by the nodes (which are neural/mental representations of stimuli) of the cognitive unit anticipates an aversive stimulus (bullying) in the presence of the auditory stimulus of the bell. **B)** When Stuart discovers that staying in the classroom avoids the hallway, he disrupts the sequence of events that leads to the aversive stimulus of bullying. Therefore, remaining in the classroom increases the probability of not experiencing bullying. As this relationship occurs, the associative trace between Class-Hallway will strengthen, increasing the probability and tranquility for Stuart that Lucas will not appear separate (Rescorla & Wagner 1972).

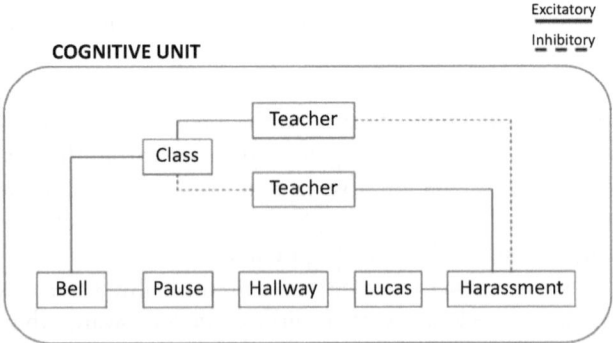

Figure 4.2 Representation of the cognitive unit of bullying 2

The presence of the teacher figure is a key element in preventing Stuart from being bullied, since the teacher is in the classroom during the break, it is a condition for not experiencing bullying. The node that neurally represents the teacher has been incorporated into the cognitive unit of bullying, which Stuart will use for decision-making.

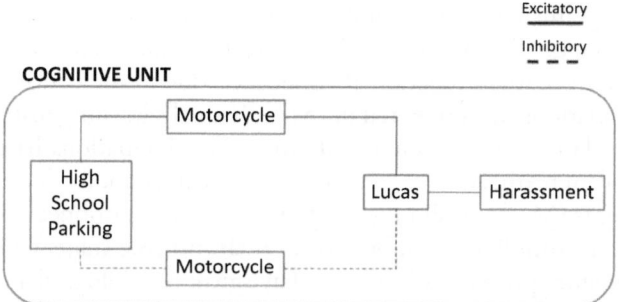

Figure 4.3 Representation of the cognitive unit of bullying 3
The presence of Lucas's motorcycle in the school parking lot anticipates whether Lucas will be there or not. The cognitive unit of bullying has changed, as finding or not finding the motorcycle at the school directly anticipates whether Lucas will be there that day, without the need to introduce other elements like the bell, hallway, etc. The nature of cognitive units is dynamic, they always change according to the stimuli perceived and their associated hedonic consequences.

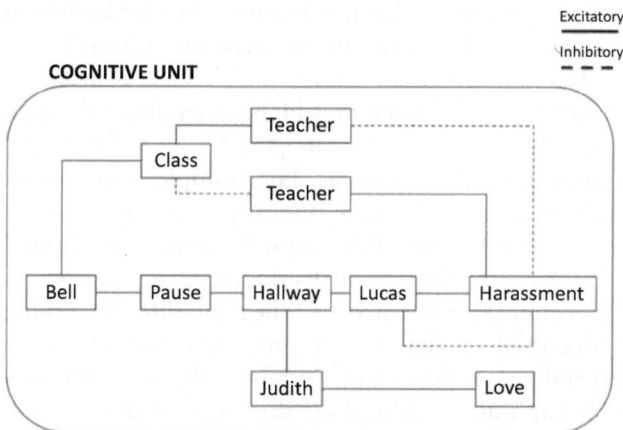

Figure 4.4 Representation of the cognitive unit of bullying 4
As new nodes are introduced into the cognitive unit, it becomes more complex and therefore requires more brain power to manipulate (León-Domínguez *et al.* 2015; Salthouse & Pink 2008; Xu *et al.* 2022). In this new cognitive unit, the consequence of bullying (aversive stimulus) competes with the consequence of love (appetitive stimulus). When two reinforcers compete, the decision-making process results in a cost-benefit analysis involving different cognitive, emotional, and contextual processes related to the value assigned by the individual to the outcomes based on past experiences (Garcia *et al.* 2023).

These diagrams represent cognitive units that people use consciously for decision-making. The nodes of these cognitive units are neural/mental representations that are functionally associated due to the contingency and contiguity relationships presented in the individual's learning history. These functional relationships enable people to make anticipations based on the detection of contextual cues (S^D) that predict the likelihood of occurrence of relevant events (Mackintosh 1975). In the previously mentioned example, all are significant stimuli, but not because of their intrinsic value, rather due to their anticipatory power of stimuli that have great physiological and psychological relevance for Stuart: harassment and love (Morrens *et al.* 2020). Each node of the cognitive unit is a simplification or reduction of the mental representation or a dimension of it to be manipulated and plan behaviors (Huang & Awh 2018). The key lies in understanding that environmental stimuli, regardless of their nature (real versus digital), automatically anticipate elaborate representations of potential consequences, which serve as a basis for planning and making decisions on voluntary behaviors aimed at specific goals: obtaining or avoiding the consequence.

Just as in Stuart's example, digital stimuli that appear on the screen of technological tools also signal different behavioral routes leading to different outcomes. Knowing the relationship between [Stimuli – Potential Behaviors – Decision Making] allows the digital behavior designer to "anticipate anticipation." This point is highly relevant, as the digital behavior designer does not analyze behaviors post hoc as a clinical psychologist would, but anticipates them a priori. For example, users have associated certain sounds with the arrival of a WhatsApp message or an email from Gmail. In this case, the value of the consequent stimulus (appetitive/aversive) is not known until the user opens the message and reads its content. The sound that anticipates the arrival of the mail, through specific mechanisms to be discussed in the chapter, may automatically trigger certain emotions depending on the context in which the messages arrive: having just experienced an unwanted breakup, the sound of the message generates hope that the content will be to re-establish the romantic relationship; having caused an accident at work, fear that the content is related to a fault or dismissal . . . To some extent, the emotion triggered by the message helps the user prepare for its content by anticipating possible consequences. How could we use this knowledge for the design of instant messaging services like WhatsApp or Telegram? How would we be "anticipating anticipation"? Simply put, what if we had the option to integrate with the antecedent stimulus (notification sound) a sensory dimension that also anticipated the value of the reinforcer? What if we gave the option to the user sending the message or email to introduce a special sound or vibration that indicated whether the content of the message is positive/negative or urgent/nonurgent? In this way, the recipient of the message would know the value of the message's content before opening it and could act accordingly.[6] The ability to anticipate the future plays an important role in animal survival. Being able to foresee threats and opportunities before they arise could lead to attempts to avoid aversive stimuli or approach appetitive stimuli (Suddendorf *et al.* 2018). Anticipating a possible future would generate an emotion congruent with the value of the consequence we are anticipating and its acquisition costs. This generated emotion, in some way, prepares you physically and psychologically for the future that is being anticipated.[7] Therefore, understanding the anticipatory nature of the brain and the emotional responses it generates is a key process in the design of digital behaviors. "Anticipating anticipation" is a central element for the design and development of ergonomic digital services with user cognition, emotion, and behavior in mind.

4.3 What Are the Theoretical Foundations of Expectations?

We can anticipate the future because, in the past, we encoded experiences into memories stored in neural clusters known as engrams (Tonegawa *et al.* 2015). That is, we recall our past experiences to manipulate our future experiences. Indeed, by evoking our past, we can remember the sequence of events that occurred and attempt to act accordingly to repeat the same outcomes from the past, modify them, or directly avoid them. The central point is to understand which cognitive unit is active in the user's working memory to anticipate their actions and emotions when using digital and virtual services. But what are cognitive units "made of"? Well, the answer depends on the philosophical approach one takes to address this point. From the perspective of digital behavior design, we understand that cognitive units are composed of a discrete relational set of nodes that are neural encodings representing an external or internal stimulus (neural representations), and that have been associated because they appeared together in the same spatiotemporal segment in the past. For instance,

---

[6] This is just an example of how the designer of digital behaviors can anticipate the users' anticipation. In this specific example, there are other variables to consider regarding the suitability of introducing this new feature for messages, such as making this option available only to certain people like spouses, children, or relatives.

[7] With "futures," reference is made to the time projection of the sequences of events necessary for the attainment of the anticipated consequences.

travel-car-road may be linked in the same cognitive unit because, typically, when one speaks or thinks of travel, there is a high probability of referring to cars and/or roads; hence, the brain understands that these elements may be functionally related (Collins & Loftus 1975). The philosophy that encompasses the principles explaining this form of understanding cognitive units is known as associationism.

Associationism was developed mainly by British philosophers during the seventeenth and eighteenth centuries. This movement posited that complex mental processes such as thought, learning, and memory could be explained through the association of mental contents into aggregates that would constitute mental representations (Tonneau 2012). This approach to the study of the mind began with Aristotle, but it was not until Thomas Hobbes in the seventeenth century that its systematic study began, to be further developed theoretically by members of the English empirical school such as John Locke, George Berkeley, David Hume, David Hartley, James Mill, and John Stuart Mill. Thomas Hobbes suggested that all knowledge is composed of relatively simple sensory impressions, whereas John Locke suggested that a human being is a tabula rasa at birth, which constructs mental representations more as the result of experiences than by reason.[8] From these ideas, an empirical methodology for the study of the mind was generated, which launched various experiments in areas such as memory and learning (Baldwin 1913). Thanks to the methodology and systematization in the study of the mind, it was discovered that associations between different elements did not occur randomly, but followed a series of natural rules or principles. These natural rules or principles formulated by the Associative School were the precursor of behaviorist psychology, mainly the ideas of conditioning, upon which they base their scientific study.

One of the first associative processes empirically, methodically, and systematically studied was classical conditioning. Classical conditioning is a type of associative learning where two or more stimuli become linked due to their contingency and contiguity relationships. This learning mechanism associates neutral stimuli, which hold no inherent value for the organism, with stimuli of high biological value. For instance, the scent of food does not have predetermined value to the organism until it is associated with a high-value stimulus like the food itself. From that

---

[8] This conception of the human being as a tabula rasa, although useful to explain the power of context in our behavior, is imprecise, since we are also determined at birth by innate behaviors and cognitive processes that will be the foundation upon which human cognition will develop (Heyes 2014).

moment on, the scent of food, previously valueless to the organism, now acquires significance, because it signals the presence of a stimulus with intrinsic value – the food. Consequently, upon smelling food, the organism anticipates that food is nearby, and activates physiological and emotional mechanisms to seek and consume it. This associative process between two stimuli, one previously without value and another with intrinsic value, is known as classical conditioning, and its most prominent figure was Ivan Petrovich Pavlov. Pavlov was a Russian physiologist who, along with Vladimir Bechterev and Ivan Sechenov, were leading figures in the Russian reflexology school, which would forever change the course of psychology. Until then, psychology had been dominated by the American psychologist William James and his introspective methodology for determining the mental states of individuals. The Russian reflexologists swept away this introspective view of scientific psychology and established the study of the observable – behavior. This group of Russian scientists believed that the motor reflex was the basic unit of behavior, and thus based their postulates on its study. Although Bechterev and Sechenov made significant contributions to the study of behavior, it was Pavlov's findings that were most influential, because he managed to formulate a set of principles that predicted animal behavior with a high probability. In fact, Pavlov did not intend to discover the principles of classical conditioning but did so serendipitously. This scientist was collecting, measuring, and analyzing the salivary response in dogs and orphaned children over extended periods.[9] Pavlov believed that saliva production was the result of a nervous reflex that occurred upon food contact with the taste buds of the tongue, which he termed the salivary reflex. Over time, the animals used in his experiments, in this case, dogs, began to familiarize themselves with the experiments and produced saliva before the food was received. In other words, the dogs learned to anticipate the food. This event caught Pavlov's attention, and he quickly understood that there must be environmental events that signaled to the dog the appearance of the food. To study this anticipation mechanism, the Russian reflexologist began to present the dog with food in the presence of other sensory elements, such as a metronome, which sounded before the dog received the food.[10] Over time, he discovered that the metronome alone was capable of eliciting the salivary

---

[9] For his work in the physiology of digestion, he would receive the Nobel Prize in Physiology or Medicine in 1904 (Berger & Rejman 2019).

[10] Pavlov's dog is globally recognized in the collective imagination by a bell. However, in his book, where he explains how he conditioned the dogs, he only makes reference to the bell once. Yet, it was the use of the bell that transcended instead of the metronome (Pavlov 1927).

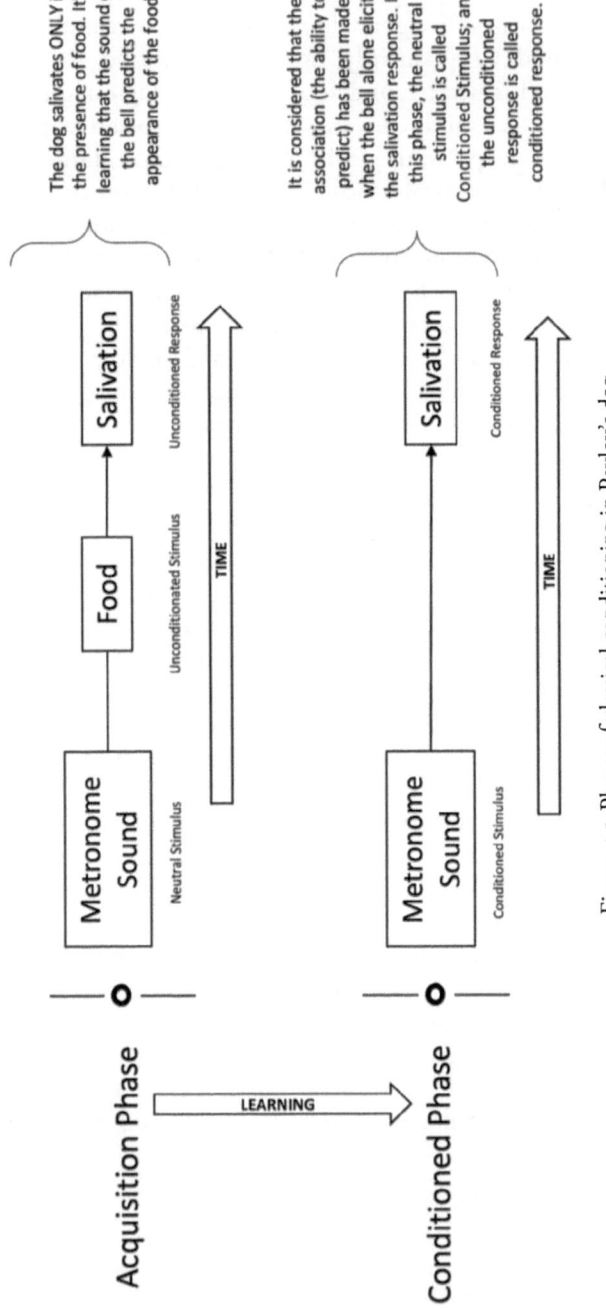

Figure 4-5 Phases of classical conditioning in Pavlov's dog

During the learning acquisition phase, the dog learns to associate a non-relevant stimulus (NS; metronome sound) with an intrinsically relevant stimulus (US; food), which elicited an automatic unconditioned reflex response (UR; salivation). The pairing of the NS with the US in the same spatio-temporal segment (contingency and contiguity relationship) produced cerebral changes where it associated that the NS preceded the US. Once the NS was capable of eliciting a UR without the need for the US to appear, it was understood that the animal had associated NS-US. In this new phase, where learning had occurred, the NS is called the CS, and the UR is called the CR.

reflex in the animal, just as if it had received the food. In a way, the metronome became associated with the food, so that the dog was able to predict its appearance and began to salivate in preparation for its receipt (see Figure 4.5). Pavlov had just discovered the basic principle that would become the main foundation of classical conditioning and one of the most fruitful psychological theories in history (Pavlov 1927).

Classical conditioning posits that organisms are capable of predicting the onset of stimuli with high biological value by associating them with a neutral stimulus, that is, a stimulus without value. In the aforementioned example, the dog learned to associate that the metronome (NS; neutral stimulus) always preceded the food (unconditioned stimulus; US), which produced a salivary reflex when ingested (unconditioned reflex; UR). When the dog was exposed enough times to this relationship, the stimulus that was previously neutral (metronome) began to acquire value for the dog because it anticipated the appearance of food. At this point, when the metronome began to generate the salivary reflex in the absence of food, it could be considered that the dog had been conditioned to the sound of the metronome. In other words, the neutral stimulus, the metronome, had sufficient power to produce the salivary reflex. In this phase, the metronome would cease to be called a neutral stimulus and would be called a conditioned stimulus, and the salivary response would cease to be called an unconditioned response to be termed a conditioned response. This change in nomenclature is used to reflect that the organism has been conditioned to a stimulus that was previously neutral. This phenomenon where the conditioned stimulus (CS) has the same power as the US to trigger the conditioned reflexes (CR) is explained through the Stimulus-Substitution Theory (Garcí-Hoz 2003), and it occurs because the CS acquires the power to release dopamine due to its association with the US (Morrens *et al.* 2020).

A very basic example that occurs in digital environments or services, but that will help to understand the idea of anticipation, occurs when using Spotify. Imagine a situation after a tiring day at work, and you leave yearning to play your favorite song while walking to public transport. You are already inside the Spotify app, about to press the play button to play your favorite song. Before you press the button, there is already a conditioned response of well-being (CR) produced by a conditioned stimulus, which in this case is the play button (CS). Your brain, upon seeing the play button (CS), anticipates that the song (US) will play, thus placing the organism already in the ideal state (well-being) to listen to it (CR). If you press the button and the song plays, that's good, everything happens as anticipated. But what happens if it doesn't play? The nonappearance of the unconditioned stimulus, when it was anticipated to appear, generates an emotional response of discomfort, which will be associated with the Spotify button, and with the Spotify brand itself.[11] The astute reader will have noticed from this example that, essentially, classical conditioning and operant conditioning are two of the same processes, and indeed, from the perspective of designing digital behaviors, this should be considered. The same process that has just been explained could perfectly replace the CS with the S^D (play bottom) and the US with the S^{C+} (song), and be explained through operant conditioning. So, how do these two learning processes converge?

4.4   Integration of Classical and Operant Conditioning in a Single System

The convergence of classical and operant learning within a unified system should be approached from the perspective that individuals are capable of generating a cognitive unit from a contextual stimulus, which contains mental representations of multiple futures that could possibly occur (Pickens & Holland 2004). The emergence of a stimulus enables a person to anticipate alternative futures by activating different mental representations that have encoded the memory of a significant past event (Colwill & Rescorla 1990). The awareness of an environmental stimulus (S^D/CS) triggers in the individual's imagination multiple behavioral alternatives to obtain various reinforcers with different values. The contemplation of these mental representations within a singular cognitive unit that can be manipulated by the individual will precede a cost-benefit analysis determining which action to perform for a specific reinforcer (Morris *et al.* 2022). This theoretical point is well elucidated by the two-process learning theory, which posits the coexistence of these two types of learning (classical and operant) and further, that they both operate under a single system (Rescorla & Solomon 1967) (see Figure 4.6).

The two-process learning theory proposes that both classical and operant conditioning are inseparable in a learning process, wherein classical conditioning provides the motivation for the operant response (Rescorla & Solomon 1967; 2006). In operant conditioning processes, the S^D

---

[11] The process by which a conditioned stimulus becomes associated with an US, without having had a functional contingency relationship, is due to its association with other CS, which did maintain a contingency relationship with the US. In the example, CS2 (Spotify), became associated with an CS1 (play button), which predicted the appearance of the US (music). In this case, since CS2 is two orders away from the US, it is known as Second Order Conditioning (Lee 2021).

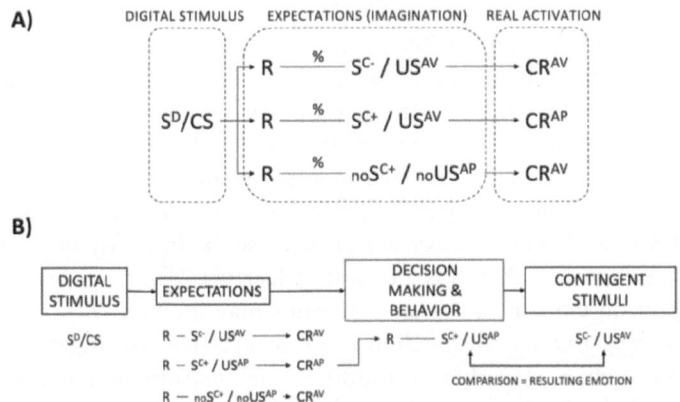

Figure 4.6 Expectations versus reality
This figure explains the emotional responses of the user when using a digital service as a result of comparing the expectations of the reinforcers that could be obtained with the reality of the reinforcer that is obtained. **A)** The perception of a natural stimulus, which may be considered an S^D or an CS, automatically activates different mental representations in the user's imagination, which are expectations of possible rewards from using the technological service (Pickens & Holland 2004). Each of these mental representations is the result of previous experiences using the digital service, the observation of another user using it, knowledge acquired instructionally, or generalizations from other previously used digital services. The user can imagine various possible behaviors to perform [R-(S^C/US)], which are associated with a probability of obtaining the S^C/US (Philiastides *et al.* 2010), and it is the mental representation of the S^C/US that causes the appetitive CR (CR^{AP}) or an aversive CR (CR^{AV}), and not the S^D/CS (Holland & Rescorla 1975; Rescorla 1973; Rescorla & Cunningham 1978). **B)** The action selection model depends on 4 phases: (1) the previously imagined representations, (2) the value of performing each action, (3) the action selection, and (4) the evaluation of the outcomes (Rescorla & Cunningham 1978). Particularly, the decision-making of action selection is achieved through a cost-benefit analysis (Basten *et al.* 2010), including the physical and mental effort to reach it (effort-based cost–benefit valuation) (Croxson *et al.* 2009). Decisions based on a cost-benefit analysis involve accepting or rejecting alternative actions due to the benefits of obtaining the reward, as in the extent to which potential punishments are avoided (Basten *et al.* 2010; Hull 1943). This process acquires special relevance in decision-making that occurs within a context of uncertainty (Philiastides *et al.* 2010). Similarly, other theories, such as Prospect Theory by Kahneman and Tversky, propose alternatives to decision-making, which would be based on how an individual envisions their future state in relation to the current subjective state (Kahneman & Tversky 2018). What is posited from the perspective of digital behavior design is that the comparison of the value of the obtained reinforcer with the value of the expected reinforcer will result in an emotion, which will become associated with the user experience of the technological service.

inherently produces a Pavlovian association, as the reinforcer appears contingent upon its presence, and does not appear when the S^D is absent. This implicit association may yield incentive-motivational effects that energize behavior (Weiss 1978). The Pavlovian association between the S^D and the reinforcer is responsible for the excitatory and inhibitory effects of the S^D on behavior (Roberts 2014; Weiss 2014). Experimentally, the proponents of this theory demonstrated in an operant paradigm how behavior was controlled by an excitatory CS [named CS(+)]. In this research, composed of three experiments, the authors trained a group of dogs on a continuously operating free-operant avoidance schedule, and showed how an CS could decrease or increase the intensity of the operant avoidance response that was previously learned. That is, the predictive capacity of an CS for an aversive stimulus may interact with the operant response by signaling the likelihood of an aversive contingency's occurrence, in such a way that it modifies the instrumental expression of avoidance behavior Separate (Rescorla & Lolordo 1965). The two-process learning theory explains the relationship of operant and Pavlovian processes through the interaction of three variables: reward expectations, the interplay of responses, and the central emotional states. Notably, changes in an individual's emotional states are subject to these Pavlovian laws, directly influencing people's motivation to perform voluntary behaviors (Rescorla & Solomon 1967). In a way, during the course of operant conditioning, the reinforcers following the operant response become associated with the antecedent stimuli, such that they are connected under the classic schema of $(CS/S^D - US/S^{C+})$ (Miller 1948). For instance, the use of Uber Eats to order food automatically triggers a sense of well-being because we anticipate receiving food (via a mental representation/cognitive units), as in the past, whenever we used Uber Eats, we received food we enjoyed. The S^D from Uber Eats also acts as an CS(+), eliciting an appetitive conditioned response (CR^{AP}). Conversely, if we had an aversive experience the last time we used Uber Eats, we anticipate that one of the possibilities is to repeat the same prior experience, hence this anticipation could trigger an emotional response of discomfort $(CR^{AV}$; aversive conditioned response), an emotion that could interfere with the decision-making process of using Uber Eats to order food again (such influences have been experimentally verified) (Estes 1948; Lovibond 1983). In its most modern version, the anticipation of what will happen is the cornerstone of human cognition. Friston proposed a revolutionary approach to understanding human cognition through the free energy principle, a theoretical framework that suggests predictive coding as the neuronal mechanism that underlies cognition (Friston 2010; Rao & Ballard 1999). This principle contends that the brain is a system of inferential computation, which is always in

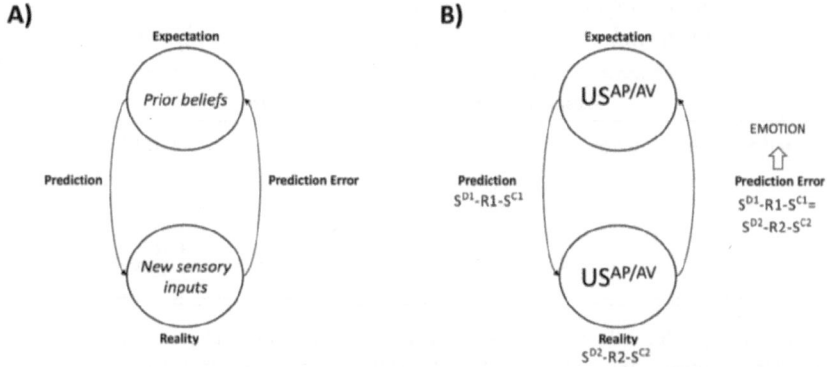

Figure 4.7 Adaptation of predictive coding theory for use in the design of digital behaviors

A) This figure illustrates the computational relationship between expectation versus reality, which is considered the foundation of learning. The brain or mind presents generative models of the world, through which they generate a prior belief or expectation in the form of a mental representation of what will occur. This expectation is compared with reality through sensory data received from the environment, thereby generating a prediction error between the expectation and reality. As the prediction error tends toward zero, it is considered that there is no learning (expectation equals reality), hence, synaptic strengthening (synaptic efficiency) occurs. Conversely, when the prediction error is significant, that is, reality is vastly different from the expectation, the generative models of the world need to be optimized so that next time they can generate more accurate predictions (prior beliefs). This optimization occurs through a "feed-forward precision-weighted signal" that updates the world's generative models, an update that is considered learning. The greater the discrepancy between expectation and reality, the greater the surprise, so the change in synaptic strength leads to a significant release of energy, hence the name of the free energy principle (Friston 2009, 2010). B) This figure is an integration of the predictive coding theory with the associations formulated by classical and operant conditioning. The expectation that is generated is related to the acquisition of an appetitive (or reinforcing) unconditioned stimulus or the avoidance of an aversive unconditioned stimulus (punishment). For the design of digital behaviors, the cognitive system generates a prediction that represents an association of the type S^{D_1}-R_1-S^{C_1}, which is subsequently compared with the outcomes obtained in reality (S^{D_2}-R_2-S^{C_2}). This comparison, ultimately, can be reduced to whether the unconditioned/reinforcing stimulus that was expected to be obtained is of equal valence and intensity to what was actually obtained. The prediction error or discrepancy is what would generate a change in the functional relationships between the different elements of the mental representation that predicts the acquisition of the reinforcer, as well as an emotional response that will become associated with the world's generative models.

continuous optimization of mental representations through the updating of probabilities associated with possible "futures" (see Figure 4.7).

This optimization is predicated on minimizing the discrepancy between prior mental representations (prior beliefs) and actual occurrences (sensory

Table 4.1 *Resulting emotion from comparing the appetitive valence of the expected stimulus with the valence of the actual stimulus received*

Expectation of an appetitive stimulus		
Expected Stimulus Valence	Actual Stimulus Valence	Resulting Emotion
S^{AP}	S^{AP}	CR^{AP}
S^{AP}	S^{AV}	CR^{AV}
S^{AP}	S^{NS}	CR^{AV}
S^{AP}	$S^{\emptyset}$	CR^{AV}

The emotions produced by the discrepancy between expectation and reality have been categorized according to James Russell's model, which encompasses the valence of emotions into two categories: pleasant (appetitive) and unpleasant (aversive) [63]. S^{AP}: appetitive stimulus (pleasant for the organism); S^{AV}: aversive stimulus (unpleasant for the organism); S^{NS}: neutral stimulus (emotionally neutral stimulus for the organism); $S^{\emptyset}$: absence of the expected stimulus; CR^{AP}: conditioned response that produces well-being; CR^{AV}: conditioned response that produces discomfort; $CR^{\emptyset}$: no emotional change.

inputs). The greater the discrepancy between expectations (prior beliefs) and reality (sensory inputs), the greater the surprise, and the more energy is required to correct the prior belief and adjust it according to the new information entering the system (Friston 2010).

When we integrate the principles of classical conditioning, operant conditioning, and the predictive coding theory, we can derive predictions about how an organism will emotionally react to events. This methodology can also be applied to robots to generate emotions. For example, Table 4.1 illustrates that if a user of a technological service (or robot) expects an appetitive stimulus (S^{AP}), it will only cause a resulting emotion of wellbeing (CR^{AP}), if the expectation and reality match in appetitive valence. That is, if I contract a service through the platform fiveer.com, it is expected that the delivered product will be pleasant (appetitive), so if the delivered product is pleasant, the user generates a pleasant emotion associated with fiveer.com. Conversely, in all other cases where a user expects an appetitive stimulus and obtains an aversive, neutral stimulus, or simply receives none, the comparison between expectation and reality will generate an aversive emotional response (CR^{AV}). That is, if an appetitive product from Fiveer is expected, and on the contrary, an aversive product is received, one that causes indifference or none is received, the resultant emotion will be aversive (CR^{AV}). This aversive emotion becomes associated with the representation of Fiveer, so that the next time you decide to use fiveer.com, this emotion may interfere with decision-making (as proposed by the two-process learning theory) (Rescorla & Solomon 1967). In cases of discrepancy, the user will always adjust their prior belief (expectation) to the new outcomes, which may influence their mental models of how the world works. Anticipating the expectations with which a user may use a technological service can help the designer of digital behaviors to optimize processes so that expectation and reality always generate an appetitive emotional response. For instance, someone might want to design a new platform to watch videos online that competes with YouTube. The primary or molar behavior of platform users is the viewing of videos. What expectations might future users have? At this stage, potential users should be asked about their usage expectations. To this end, the digital behavior designer must first create space to query the users, and have a set of questions prepared regarding potential appetitive and aversive stimuli, where the emotional intensity that would be generated by encountering them during the use of the digital service will be indicated. For example, before the interview with potential users, the digital behavior designer may anticipate that the appetitive stimuli they expect to encounter during the use of an audio-visual platform are: the quantity and variety of videos, ease of finding videos, loading time, image and sound resolution, the ability to down-load videos/audios . . . any of these unmet expectations will automatically trigger an aversive emotion that will modify the user's mental represen-tation of the digital service. In this example, it may be that videos take a long time to load because the developers did not pay attention to image compression systems. As mentioned in previous chapters, the delay in video loading can reduce the value of the video and the use of the platform because the expectation of quickly viewing them is not met, generating an aversive emotional response. This discrepancy between expectation and reality causes the mental representation of the platform to be adjusted with new information such as "videos on this platform take time to load." Therefore, whenever there is a possibility of viewing a video on this platform, the user will automatically activate a cognitive unit that will be used for decision-making, in which there will be a node that has encoded an aversive emotion due to the slow loading of the videos. If this platform were the only one on the market (there are no other potential behaviors that could be performed to view the video), it could happen that the user gives up on watching the video or watches it for some other associated value of the video. The problem arises when there are other platforms that have anticipated user expectations and have paid attention to these elements during the design and development of the service. Or if there is no competition yet, this is a weakness that can motivate other entrepreneurs to exploit it and develop a new service that competes with this platform.

The foregoing discussion addressed what happens when user expectations are pleasant in the use of the digital service. Subsequently, we will show what occurs when the expectation is aversive, neutral, or absent (see Tables 4.2–4.4).

In the development of new digital services, user expectations are commonly anticipated to be appetitive. However, on many occasions,

Table 4.2 *Emotion resulting from comparing the aversive valence of the expected stimulus with the valence of the actual stimulus received*

Expectation of an aversive stimulus		
Expected Stimulus Valence	Actual Stimulus Valence	Resulting Emotion
S^{AV}	S^{AP}	CR^{AP}
S^{AV}	S^{AV}	CR^{AV}
S^{AV}	S^{NS}	CR^{AP}
S^{AV}	$S^{\emptyset}$	CR^{AP}

S^{AP}: appetitive stimulus (pleasant for the organism); S^{AV}: aversive stimulus (unpleasant for the organism); S^{NS}: neutral stimulus (emotionally neutral stimulus for the organism); $S^{\emptyset}$: absence of the expected stimulus; CR^{AP}: conditioned response that produces well-being; CR^{AV}: conditioned response that produces discomfort; $CR^{\emptyset}$: no emotional change.

Table 4.3 *Emotion resulting from comparing the neutral valence of the expected stimulus with the valence of the actual stimulus received*

Expectation of a neutral stimulus		
Expected Stimulus Valence	Actual Stimulus Valence	Resulting Emotion
S^{NS}	S^{AP}	CR^{AP}
S^{NS}	S^{AV}	CR^{AV}
S^{NS}	S^{NS}	$CR^{\emptyset}$
S^{NS}	$S^{\emptyset}$	$CR^{\emptyset}/ CR^{AV}$

S^{AP}: appetitive stimulus (pleasant for the organism); S^{AV}: aversive stimulus (unpleasant for the organism); S^{NS}: neutral stimulus (emotionally neutral stimulus for the organism); $S^{\emptyset}$: absence of the expected stimulus; CR^{AP}: conditioned response that produces well-being; CR^{AV}: conditioned response that produces discomfort; $CR^{\emptyset}$: no emotional change.

Table 4.4 *Emotion resulting from comparing the absence of expected stimulus valence with the valence of the actual stimulus received*

Expectation of the absence of a relevant stimulus		
Expected Stimulus Valence	Actual Stimulus Valence	Resulting Emotion
$S^{\varnothing}$	S^{AP}	CR^{AP}
$S^{\varnothing}$	S^{AV}	CR^{AV}
$S^{\varnothing}$	S^{NS}	$CR^{\varnothing}$
$S^{\varnothing}$	$S^{\varnothing}$	$CR^{\varnothing}$

S^{AP}: appetitive stimulus (pleasant for the organism); S^{AV}: aversive stimulus (unpleasant for the organism); S^{NS}: neutral stimulus (emotionally neutral stimulus for the organism); $S^{\varnothing}$: absence of the expected stimulus; CR^{AP}: conditioned response that produces well-being; CR^{AV}: conditioned response that produces discomfort; $CR^{\varnothing}$: no emotional change.

particularly when a technological service has failed and the company invests in its revival, the designer must foresee user expectations as aversive, focusing on those elements that can shift an aversive expectation to an appetitive one. Take, for instance, the previous example of the video platform: imagine it has been operational for five months without gaining significant user traction. During a technological audit, a digital behavior designer discovers that video uploads are excessively slow due to the videos' large file size. It is explained that users experience a phenomenon known as "delay discounting," which leads to the diminution of the reinforcer's value due to the delay in its delivery. After a meeting with the development team, they understand the problem and manage to implement an algorithm that reduces video sizes, thereby increasing upload speed. Consider Nicole, a former user of the platform, who must return to it because it exclusively hosts a video she wishes to watch. She approaches the platform with a negative expectation (S^{AV}). When Nicole clicks play, she finds that the video loads promptly (S^{AP}), an event that creates a discrepancy between expectation and reality, eliciting a sensation of well-being (CR^{AP}) (see Table 4.2). This sense of well-being reconfigures Nicole's mental schema of the platform, possibly enabling the recognition of other reinforcers that were previously obscured due to a blocking effect caused by her adverse experiences. She might not have noticed the high quality of the videos or the efficient search and recommendation system because of the negative experience from the slow video loading times. In other words, aversive emotions linked to a service due to any discrepancy between expectation and reality can overshadow the effect of other well-designed reinforcers. Another example of utilizing user expectations to design favorable elements for digital services is when there is no relevant pre-existing expectation during an interaction with a digital service. Take food delivery apps, for example. When a user clicks the payment button, they may expect nothing out of the ordinary ($S^{\emptyset}$), merely to confirm the payment. But what if a pop-up appeared offering an unexpected 5 percent discount? Clearly, this appetitive stimulus, which could also be considered a reinforcer, would generate a sensation of well-being associated with the digital service. What if we introduced random discounts to users under fixed-ratio and fixed-interval conjunctive reinforcement schedules? That is, if a user places, for instance, three food orders within a week, they are automatically granted a discount on their next order without prior notice. Once the habit is established, the requirement could be progressively increased from three to four, and from four to five orders, to tailor the offer to the financial models of the business. This is just one instance of how knowledge from behavioral and cognitive sciences can be leveraged for the design of digital behaviors.

CHAPTER 5

Our Internal States As a Source of Motivation

5.1   Internal Milieu or Internal State

Individuals are aware of their desires in the realm of the familiar, yet they remain unaware of their desires stemming from the unknown. Possessing a theoretical understanding of the potential needs users may encounter, enables a digital behavior designer to suggest various digital service features that address these needs. The digital behavior designer directs the user's motivation to use the digital service through their design. In less scholarly or popular psychology texts, the study of motivation has been approached from a flawed conceptualization, often conflating the concept of "reinforcer" with motivation. While both concepts are closely related, they are not synonymous. To grasp this distinction, one must understand that the organism always operates with a specific internal state or milieu intérieur, which generally demands resource consumption. These resources tend to be environmental stimuli with chemical and/or representational attributes potent enough to satisfy the organism's internal demands. For instance, food quells hunger (a chemical attribute), whereas a smile can satiate an internal psychological state of uncertainty (a representational attribute). Specifically, the anticipation of these attributes of the satiating stimulus (or reinforcer), when the organism is experiencing a state of internal deficiency, confers value to the stimulus, thereby originating and controlling behavior.[1] Once the reinforcer is obtained and consumed, it satiates the

---

[1] A potential explanation for what sustains the energization of exploratory behavior may be offered by the imbalance in caloric deficit and the resultant hormones released from this deficit (Broberger 2005). In Thorndike's original experiments with puzzle-boxes, cats were deprived of food to motivate them to attempt to escape the box and obtain the food reinforcer. This methodology of depriving/ satiating with food was replicated and operationalized by Skinner in his experiments with rats and pigeons, as he considered it a motivational operation to place the animals in an optimal state for performing the target behaviors (Laraway *et al.* 2014). In states of deprivation of the reinforcer, its value increases, thus animals are more motivated to initiate exploratory search behaviors. Conversely, in states of satiety of the reinforcer, animals tend to inhibit exploratory search behaviors.

state of deficiency, causing its value to fade. In the aforementioned example, when an organism is hungry, it is the value of the chemical attributes of food that truly originates and controls the behavior of ordering a dish via Uber Eats, not the food per se. If it were the food that originated and controlled behavior, then every time we thought of or saw food, we would perform actions to obtain it. Thus, it is the internal state of nutrient deficiency that bestows value upon the food, and it is this value for restoring balance that originates and controls behavior. In the field of digital behavior design, the term "goal-directed behavior" is recommended over motivation. Goal-directed behavior is the interplay between a specific internal state, the anticipation of the reinforcer's value, and the initiation of a sequence of behaviors aimed at seeking, obtaining, and consuming the reinforcer to satisfy an internal state. Once the reinforcer is consumed, goal-directed behavior fades, and a new internal state generates different needs to be addressed. Therefore, in behavior, there is a preceding bidding element which are the functional internal states of individuals, signaling the value of environmental stimuli as the cause of goal-directed behavior.

5.2   Need, Goal-Directed Behavior, and Reinforcer

In popular psychology, and even within scientific psychology, although we use need and motivation in separate statements, when asked to define these concepts, their boundaries blur and converge. In digital behavior design, it is essential to differentiate these elements as they explain the reasons behind user behaviors. To begin with, and pending a more in-depth discussion, the relationship among these elements is sequential. That is, first, the user experiences a need generated by a change in their internal state. The internal state always tends toward equilibrium, so when this imbalance cannot be restored by the system of feedback controls, which stabilizes and normalizes the physiological capacity of the organism, it creates a "biochemical broth" of hormones and molecules with sufficient potential to energize the individual and initiate behaviors to obtain the necessary stimuli to restore balance (Stice *et al.* 2013). At this moment of internal deficit, the individual finds themselves in an energized state but does not know where to direct the energy. It is the visualization or anticipation of the stimulus, with characteristics capable of restoring internal equilibrium, that heightens brain activity in areas associated with attention, memory, and motivation to direct the "internal energy" and initiate goal-directed behavior (Stice *et al.* 2013). It is important to understand at this point that a digital service user does not act because they will achieve a reinforcer, but because obtaining the reinforcer helps to satisfy a need (Hull 1943). The sequence [NEED → ANTICIPATION OF REINFORCER → GOAL-DIRECTED BEHAVIOR → OBTAINING REINFORCER] is not a perfect formula, but it is valid to explain the series of events that occur from a state of internal need to its satiation.

In digital behavior design, the need is operationalized as an internal state, which decisively affects the perception and magnitude of the reinforcer's value and, therefore, the urgency that will lead the user to use a digital service to obtain it. Considering the interaction of the organism's internal factors with external factors allows the digital behavior designer to understand the subjective value of reinforcers that have sufficient power to change or satisfy the individual's internal states (Minamimoto *et al.* 2009). It follows that the value of a reinforcer is not solely defined by its intrinsic (hedonic) and sensory (nonhedonic) characteristics, but also by its subjective value, which is determined by the internal state of the subject. For instance, when an individual experiences hunger (an internal state), their cognitive processes simulate various plans (opening Uber Eats, looking for a place to eat on the street, calling a friend to dine together, etc.) to obtain the reinforcer (food). Each of these potential plans is manipulated collectively as a single cognitive unit to which a set of cognitions about their possible unfolding is applied. This generates an emotional subjective experience that assists the individual in deciding which is the "best" plan to secure the reinforcer, taking into account the contextual conditions (Bechara & Damasio 2005). Imagine, once again, that a user is hungry and orders a pizza (reinforcer) through Uber Eats. After about thirty-five minutes, a delicious pizza arrives, satisfying your hunger. The satiation of hunger is explained by a change in the internal state that deactivates the caloric deficit due to lack of food, thereby also reducing the energy driving the motivated behavior to find food. With the internal state changed, other needs emerge in the person's hierarchy of needs, while the need for nourishment is considered satisfied and descends in the order of priorities. These emerging needs at the top of the hierarchy as priorities bring with them the necessity to obtain new reinforcers, which will be signaled by new discriminative stimuli. In other words, once a need is fulfilled, the value of the surrounding environmental stimuli is reconfigured in such a way that it allows the identification of those new reinforcers that can satisfy the emergent need (see Figure 5.1).

The internal state is contingent upon the functional physiological and psychological state of the users, which induces emotional and behavioral changes (Flavell *et al.* 2022) due to the response of the amygdala and

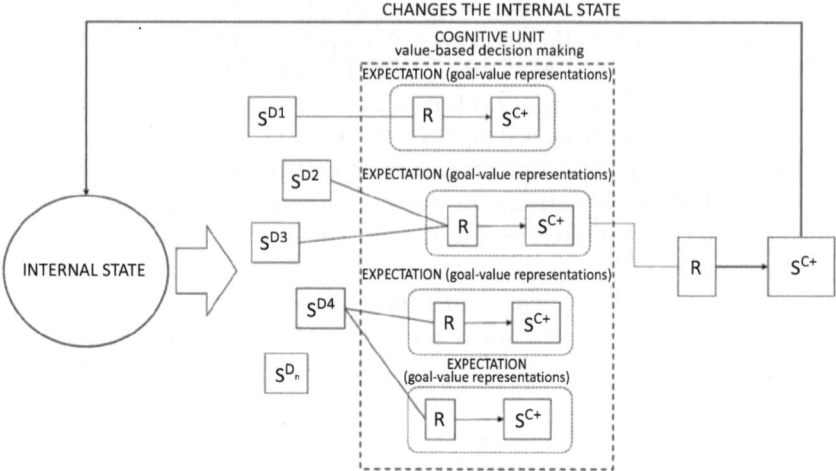

Figure 5.1 Reinforcer selection to sate internal states of the user

A user engages with a digital service to address a deficit in internal state, which may be either physiological or psychological. This psychological or physiological deficit modulates attentional processes in such a way that it detects different environmental discriminative stimuli (S^{Ds}) that signal the likelihood of obtaining a reinforcer according to the needs of the internal state (Stice *et al.* 2013). These S^{Ds} activate expectations, which are associations between an operant response (R) and a consequent (S^C). A single discriminative stimulus may signal one expectation [R-S^C] or two different expectations [R-S^C]s. For instance, an S^{D1} might be car keys, which indicate that driving to your favorite restaurant (R) will allow you to order your favorite dish (S^{C+}). Conversely, S^{D2} (Uber Eats app icon) and S^{D3} (Uber Eats website) are two different discriminative stimuli, yet both suggest that if you engage in the behavior of ordering food through Uber Eats (R), it will be delivered to your home (S^{C+}) (known as convergent multiple control). Lastly, a single S^{D4} (refrigerator) may signal two different [R-S^C], where one is preparing (R) a sandwich (S^{C+}), or eating (R) yesterday's leftovers (S^{C+}) (known as divergent multiple control). The user considers all these options and conducts a cost-benefit analysis to select the [R-S^{C+}] with the highest subjective value for them. Once the behavior is performed and the reinforcer is obtained, the chemical and/or representational properties change the internal milieu, hence altering the needs, and consequently, the value of the reinforcers and the S^D that signal them. In other words, attention is reconfigured to detect other S^{Ds} in the environment (O'Doherty *et al.* 2017).

hypothalamus (Gründemann *et al.* 2019; Lovett-Barron *et al.* 2017; Xu *et al.* 2020). The internal state that is formed as a response to the activity of brain structures may vary in intensity and topography over time, and transition into other states at different life stages as the individual satisfies their needs through the acquisition of reinforcers (Anderson 2016; Tinbergen 1951). These neuronal mechanisms that control and direct an individual's behavior, thereby modifying their internal state, appear to have an ancient phylogenetic origin, as they are present in a wide variety of animals, from those with highly complex nervous systems such as humans to those with basic nervous systems like jellyfish (Weissbourd *et al.* 2021). In other words, the fluctuation of internal states over time is nature's solution for setting organisms in motion. These shifts in organisms' internal states would be characterized by their genetic configuration which determines the physiological and psychological fluctuations, the persistence in achieving reinforcers, the topography of the behavior, the escalation in the intensity of the behavior, the generalization of internal states to other contexts, and the emotion associated with the internal state (Flavell *et al.* 2022). In digital behavior, an internal state of a user is understood as a state of the organism caused by some physiological and psychological imbalance that needs restoration.[2] Extensive literature has addressed this topic, and although all branches of psychology acknowledge the importance of internal states in modulating behavior, they do not all agree on its definition. In digital behavior, we will approach the phenomenon of describing the internal state from the conceptualization made by Clark Hull, an early twentieth-century American psychologist who attempted to integrate findings from environmental behavioral sciences with neurological findings, leaving open a space for cognitive processes when explaining the mediation of glandular and brain activation in behavior (Hull 1943). Hull posited that needs are a source of motivation for organisms that operate both in habit formation and in their functioning. Needs exert a sensitizing or dynamizing action on motivation, which he will term "drive."[3]

It seems Hull refers to drives in two senses: (1) as a neuronal system that controls an internal physiological state regulated by hormonal secretions induced by internal glands with various causal origins, or (2) a state of the animal wherein the behavioral system is activated, an activation equated with motivation (Baerends 1976). In both cases, he names the energy that

---

[2] In the design of digital behavior, allostasis can be considered an active process that enables the maintenance or reestablishment of homeostasis. In this context, allostasis refers to the body's ability to produce hormones such as cortisol and adrenaline, as well as other mediators, for example, cytokines and parasympathetic activity, which assist the animal in adapting to new situations or challenges, both predictable and unpredictable. Nevertheless, some experts prefer to use the term homeostasis to describe what allostasis accounts for. Here, homeostasis encompasses the physiological aspects that keep us alive, as well as those that aid us in adaptation (McEwen & Wingfield 2010). For a more in-depth review of the discrepancies between allostasis and homeostasis, see (Ramsay & Woods 2014).

[3] With the aim of generating our own categorical closure, the term "drive" will be used in place of "impulse."

causes the activation of these glands "stimulus drive," commonly known as motivation. A thorough reading of his book does not make entirely clear the distinction between need and motivation, as he often uses the term drive to refer to both interchangeably. Indeed, he even likened it, in a sense, to the Freudian libido (Hull 1943). This issue was already noted by Skinner in his book "The Behavior of Organisms" when he said: "At one extreme, the drive is considered simply the basic energy available for the organism's responses; at the other, it is identified with the purpose or some internal representation of a goal." Taking Skinner's interpretation of drive as a reservoir of available energy, and considering Hull's definition of drive as a general sensitizer that increased the strength of the response (Hull 1938), for digital behavior, the drive is an internal vector that signals the direction of the energization of behavior, but does not energize it. This energization of behavior would be defined by the underlying emotion generated by the anticipation of reinforcement schedules. This subject will be discussed in depth later on. For now, we shall start by defining and operationalizing drive as an internal vector that signals the direction of behavior. Therefore, to recapitulate, in the design of digital behavior, the drive is the need and the motivation is the underlying emotion. That is, the drive places the individual in a moment of need that must be resolved, while the emotion signals the energy or impetus with which the individual will display motivated behavior to obtain a reinforcing stimulus to satiate the drive.[4] Finally, in a context where there is a drive and an emotion that energizes the subject, the discriminative stimulus will initiate goal-directed behavior by signaling the organism's opportunity to obtain a reinforcer.

Drive, within the context of digital behavior, is defined as a need. Throughout the text, we will refer to it as "drive" to establish a unique jargon distinguishing digital conduct from other theoretical and practical approaches. Drives can be regarded as an internal state triggered by hormonal or neural changes that may lead to physiological and/or psycho-logical imbalance. An individual becomes aware of their drive when signaled by an internal stimulus (like the pang of hunger) or an external one (such as a threatening gaze indicating danger), eliciting an emotional response (stemming from anticipation) that seeks satiation through the acquisition of a reinforcer. For instance, when a user experiences a hunger

---

[4] It has been observed that it is not always necessary to directly address the problem presented in the drive, as there are certain behaviors that can temporarily mask it, such as the performance of pleasurable and entertaining activities. For instance, it has been studied that social-type activities can release oxytocin, which could suppress appetite (Lawson *et al.* 2015) or anxiety (Churchland & Winkielman 2012).

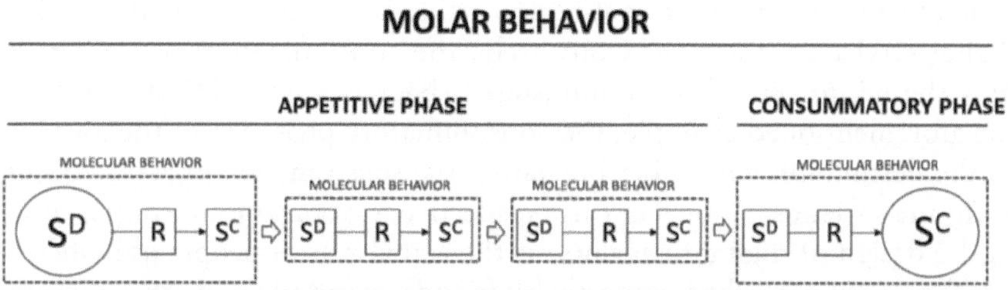

MOLAR BEHAVIOR

Figure 5.2 Interpretation of digital behavior as instinctive conduct
The appetitive and consummatory phases would together constitute what would be molar behavior in the design of digital behaviors. The appetitive phase would be composed of a concatenation of molecular behaviors that would have to occur until the molar reinforcer is obtained. The attainment or sequence of the ultimate molecular $[S^D\text{-}R\text{-}S^C]$ of the molar behavior would coincide with the consummatory phase, which can also be designed.

drive and orders food via Uber Eats, the entire sequence of motor actions culminating in pressing the order button can be viewed as motivated behavior. This motivated behavior can be divided into two phases: an exploratory or appetitive phase, eventually leading to a consummatory phase of the reinforcer. In the study of animal behavior, these phases are also referred to as instinctive behavior.[5] In the design of digital behaviors, the appetitive phase encompasses a number of molecular behaviors, which involve the chaining of molecular reinforcers until the acquisition of the molar reinforcer (see Figure 5.2).

Another example involves the use of the Uber app. A person aiming to fulfill a social contact drive might decide to meet with friends downtown. Knowing they will consume alcohol, they opt to request an Uber. Opening the app, the display of fare prices, the notification of the car en route, its arrival at one's home, and finally reaching the downtown meeting spot – each of these are molecular reinforcers within the appetitive phase that guide one to the molar reinforcer, which in this case is the interaction with friends. It is crucial to note that these molecular reinforcers adhere to the same principles as the molar reinforcers and are similarly influenced by their mode of delivery. For example, a delay in the Uber's arrival could

---

[5] This instinctive behavior is triggered and controlled by releasing or discriminative stimuli that signal the opportunity to obtain stimuli with certain value to satiate the drive present in the individual (Miller *et al.* 1960). In animal studies, Konrad Lorenz introduced the concept of "innate releasing mechanisms" to describe the fact that organisms exhibit a fixed action pattern in response to a certain releasing stimulus (Williams 2019).

generate an aversive emotional response that becomes associated with the digital service in question. Conversely, the consummatory phase details how the reinforcer will be administered (Baerends 1976; Hinde 1953). In the aforementioned example, the consummatory phase is not the journey to the destination with Uber but rather the social interactions that occur with one's friends. Adopting this perspective is of paramount importance in the design of digital behaviors, as the actual reason people use Uber is not for transit per se but to meet with friends, commute to work, or return home, among others. Typically, the focus is on ensuring a pleasant ride – clean cars, adherence to agreed fares, courteous drivers – but the real need or drive prompting the user to choose Uber is not considered. When this viewpoint is adopted, services begin to be seen differently. What if, when we requested an Uber, we also specified the reason? What if each reason for using Uber was linked to a reinforcement schedule offering points redeemable for food and beverages at various bars and restaurants? This is merely one illustration, as identifying the drive and its consummatory phase could lead to other propositions, such as offering a drink for the journey when meeting friends. Tailoring the service to the drive can have an impact on more rapid habit formation and differentiation from competitors. Hence, drives are at the origin of motivated behavior in organisms, placing individuals in an internal functional state that triggers an appetitive behavior, which will only be satisfied by obtaining the appropriate reinforcing stimulus. It is incumbent upon the designer of digital behaviors to anticipate the drives with which users approach digital services and to design the digital pathway for the attainment of the reinforcer that will satiate or aid in ultimately satisfying the drive.

5.3   Motivated Behavior Is Understood As Goal-Directed Behavior

Previously, the term "motivated behavior" was primarily used to express the concept of a behavior seeking the satiation of a drive. This term has been employed to facilitate the reader's comprehension and to introduce other complex terms such as drive. Indeed, from the perspective of cognitive sciences and the design of digital behaviors, motivated behaviors are defined as goal-directed behaviors. When Skinner's operant behavior model is integrated with Hull's concept of drive as an underlying element of voluntary behavior in organisms, it is referred to as a goal-directed control behavior. Goal-directed control behavior is a class of operant response that is controlled by the specific outcomes thereof, that is, the goal. By incorporating the term "goal" into the concept, we empower the understanding that

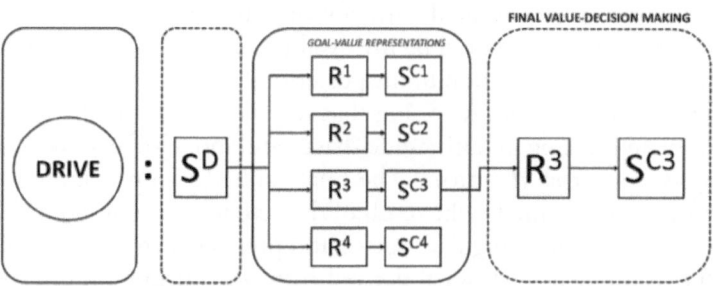

Figure 5.3 Explanation of goal-directed behaviors
Goal-directed behaviors are based on the consideration of different actions with various
goal-value representations for executing a final value-decision making process to reach
a determined goal. Goal-directed behaviors are the result of integrating within a single
theoretical framework Hull's drives, Skinner's operant conditioning, and model-based
learning processes.

voluntary behavior, in our case digital, often seeks to achieve results, which
may have a reinforcing value for the performed behavior. The integration of
this entire set of expectations with the individual's impulses forms a flexible
control system that allows the organism to project environmental contingencies before obtaining them. This goal-oriented behavior control system
has been found in both humans and other animals (de Wit *et al.* 2009;
Dickinson 1985; Dickinson & Balleine 1994). Following the emergence of
a drive, individuals calculate a representation of the expected value or utility
of a satiating stimulus (goal-value representations) that is contingent on the
selection of a particular option (see Figure 5.3).

This calculation facilitates a comparison process that enables the identification and execution of the option with a higher expected value
(O'Doherty *et al.* 2017). Daw and colleagues propose that individuals
encode an internal model on goal-value representations, which involves
projecting into the future the different states and relevant actions, as well as
the transitions between them (Daw *et al.* 2005). Individuals use this map of
goal and value representations to make flexible, real-time calculations
when considering the anticipated value of the outcomes of their potential
behaviors (expectation), and to the integration of these expectations
with the knowledge of how to perform them. This type of cognitive
model-driven reinforcement learning process is known in computational

sciences and cognitive neurosciences as model-based reinforcement learning (Fürnkranz *et al.* 2011). Since model-based action values depend on arithmetic computations that explain quantity and probability, brain systems traditionally associated with working memory, such as the dorsolateral prefrontal cortex, could be involved (Miller & Cohen 2001). These prefrontal regions, in addition to the posterior parietal cortex, are the brain areas related to neuronal processing that results in states prediction error, which explain learning in both model-based and model-free reinforcement learning (Gläscher *et al.* 2010).[6] The final value-decision making for the behavior to be performed is the result of the activity of the aforementioned brain regions (anterior cingulate cortex and prefrontal cortex) that encode the value of the neuronal signal derived from the integration of expected benefits and costs of potential behaviors (Hosokawa *et al.* 2013; Hunt *et al.* 2012).[7] The conception of behavior as a goal-directed control system, whose objective is to satiate an emerging drive from an imbalance in the internal state of a potential user, is the fundamental pillar from which user behavior in digital services will be analyzed and anticipated.

5.4   Drives As Control of Behavior

In the seminal work of Clark Leonard Hull from 1943, entitled "Principles of Behavior: An Introduction to Behavior Theory," a significant effort is made to develop a scientific system within psychology to articulate 16 fundamental principles of behavior, later known as "Hull's Principles of Behavior" (Hull 1943). Although this book has elicited considerable critique, particularly concerning its neurological analysis of behavior, it contains segments deemed to possess substantial intellectual value in analyzing and understanding human behavior (Hull 1943). The central

---

[6] In the mammalian brain, it is supposed that there exist two learning systems that guide behavior, which are also used for decision-making: model-free and model-based learning processes (Doody *et al.* 2022). Model-free learning encodes the value of behavior through reward prediction errors (Dayan & Niv 2008); model-based learning encodes the value of behavior through projecting the values of future states (through cognitive maps of the world) rather than the consequences of immediate actions, thus being connected to the environment and internal states (Gläscher *et al.* 2010). Both systems seem to be interconnected, though it is a debate that remains open (Doody *et al.* 2022).

[7] These investigations point to the possibility that the valuation of behavior involves the interaction between multiple brain systems, and that goal-value representations in the ventromedial regions of the prefrontal cortex are ultimately integrated with action information in the dorsal cortical regions to compute an overall action value (O'Doherty *et al.* 2017). Once the organism is prepared to perform the behavior, it is model-based planning, in such a way that it utilizes memories and experiences previously stored to execute the operant response (Pfeiffer & Foster 2013).

aim of the book was "to elaborate basic molar behavioral laws that underlie the social sciences." In other words, Hull sought to comprehend how and why individuals behave within a sociocultural environment. To answer this, Hull formulated the drive reduction theory (Hull 1943). This theory posited that the survival of a species' individuals hinges on their physiological state being in perfect condition; thus, any imbalance would give rise to certain primary needs. These primary needs would manifest due to the activation of specific internal glands, energizing the organism to initiate behaviors that would restore the previous internal equilibrium. This concept is particularly salient for the design of digital behavior, temporarily setting aside the neural explanation of the phenomenon. Hull exemplified with the drive generated by caloric deficit, commonly referred to as "hunger." The urgency of resolving this primary drive could be influenced by preceding conditions such as the duration of prior food deprivation.[8] Hull noted that the biochemical imbalance causing hunger, thirst, drowning, pain, cooling, suffocation, constipation, urinary retention, exhaustion (both physical and mental), insomnia, sexual abstinence, and inactivity are primary drives. When any of these drives present in an organism and a potential reinforcer is anticipated, it generates goal-directed behavior, which is modulated by the individual's emotional state (Hull 1943) (see Figure 5.4).

Primary drives were conceptualized as fundamental to the preservation of the species (Hull 1943). Therefore, the drive reduction theory suggests that goal-directed behaviors are determined by the urgency to resolve a drive, initiating a behavior in the organism that seeks to obtain the appropriate reinforcer to satiate and reduce the drive. And it is this satiation of the drive that is the primary cause of the reinforcement process (Bechterev 1932; Cannon 1932; Finan & Taylor 1940; Keramati & Gutkin 2014) (see Figure 5.5).

Let us now analyze a case of motivated behavior through a digital service available on all smartphones, the alarm clock. After a strenuous morning at work, a user feels tired (exhaustion drive), energizing the individual to initiate a goal-directed behavior to take a rest. Moreover, it is crucial that he rest due to an impending important presentation before the executive

---

[8] Hull's proposal was to encourage the mathematical quantification of behavior through the development of equations that would express the level of drive, in this specific case of hunger, as a function of the time elapsed since the last food intake (valued in hours). Furthermore, Hull proposed the creation of a second equation that would relate the intensity of the organic response with the degree of stimulus drive or motivation, which would somehow be combined with the strength of the acquired habit (Hull 1943).

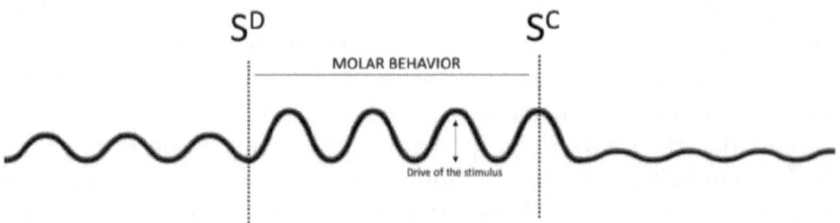

Figure 5.4 Effects of SD and SC on the drive of organisms
This wave with changes in its amplitude represents the effects of environmental stimuli on the drive of a potential user. The user has accumulated a stimulus drive, which fluctuates over time. Upon encountering an S^D that signals the possibility of obtaining an S^C that satisfies their drive, their stimulus drive or internal energization increases, triggering different neuronal mechanisms that facilitate the emission of molar behavior. Once the S^C is obtained and consumed, the stimulus drive decreases, diminishing the urgency to resolve the drive, and giving way to other drives with higher accumulated stimulus drive.

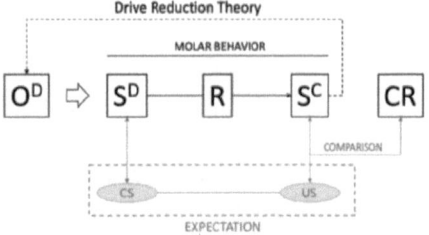

Figure 5.5 Diagram for the functional analysis of digital behaviors
This figure integrates all the elements necessary for the analysis and design of potential behaviors that a user may exhibit when using a technological service. The user starts from an internal state or drive (O^D), which biases cognitive resources to find stimuli indicating potential reinforcers that satisfy the drive. When the user encounters an S^D and decides to perform a behavior (in this case, digital), an emotional response triggered by the anticipation of obtaining the reinforcer ensues. The physical/chemical or representational attributes of the reinforcer will satiate the organism's drive (O^D) (Drive Reduction Theory), while also generating an emotional response that will be compared with the expected one to emotionally evaluate the use of the service, an emotion that will be linked to it (Hull 1943).

committee, thus having covert verbalizations (expectations) such as "if I am rested, I will be able to present and respond to questions better in my presentation." These self-verbalizations evoke an emotion of stress represented by high arousal and a negative valence of the exhaustion drive. The person considers multiple goal-value representations (plans), such as going home to nap, but time is insufficient for a round trip. Therefore, he begin an exploratory behavior of where to rest at work. Upon this behavior, they find an empty office with a bolt, thus making a final value-decision to take a short twenty -minute nap with the door bolted. Not wanting to risk oversleeping, they decide to set an alarm on their smartphone. The user proceeds with the rest until the alarm rings, awakening them. After waking up and a brief period of grogginess, they feel refreshed and energized for the afternoon presentation, which was successful as anticipated. What has occurred here? Primarily, there are several behaviors to analyze, but we will focus on the digital behavior of setting the alarm. According to the principles of operant behavior, it is highly probable that the user will reuse the office and alarm in a situation involving an exhaustion drive at work with limited time. According to Skinner, it is the sound of the alarm after twenty minutes (positive reinforcement) that selects the aforementioned behavior, but in the context of digital behavior design, it is the reduction of the exhaustion drive that actually selects the behavior of taking a nap and setting the alarm at work, not the sound of the alarm. In essence, it is the elimination of the sensation of tiredness that selects the conducted behavior. As a result, a declarative episodic memory is formed,[9] which can be accessed in similar situations as a goal-value representation (see Figure 5.6).

In the aforementioned example, the expectations formulated as covert verbalizations regarding the importance of rest, and the context (the implicit social rule that sleeping at work is frowned upon) determined the use of an office with a bolt and an alarm. However, it could also be the case that, given the same context or drive, the behavior's topography might have been different. For instance, opting for a coffee, given that there was no time for rest due to another meeting prior to the presentation. To reach the reinforcer that would allow the elimination of exhaustion for the afternoon presentation, the individual deployed various cognitive tools that enabled the anticipation of different reinforcers, decision-making, and the planning and motor execution of the chosen behavior. Hull discussed this point and noted that the variability, inconsistency, and specific unpredictability of motor responses under seemingly constant conditions are universal characteristics of organism

---

[9] According to Tulving, declarative memory can be subdivided into two categories: semantic memory, which stores information about facts and general knowledge of the world, and episodic memory, which allows an individual to re-experience a past event in the context in which it originally occurred (Tulving 1983). The ability to selectively encode and retrieve past events is an adaptive skill found in episodic memory (Castel 2007). It has been suggested that reinforcers play a crucial role in the modulation of episodic memory, enhancing the encoding of information that anticipates reinforcers (Mason *et al.* 2017).

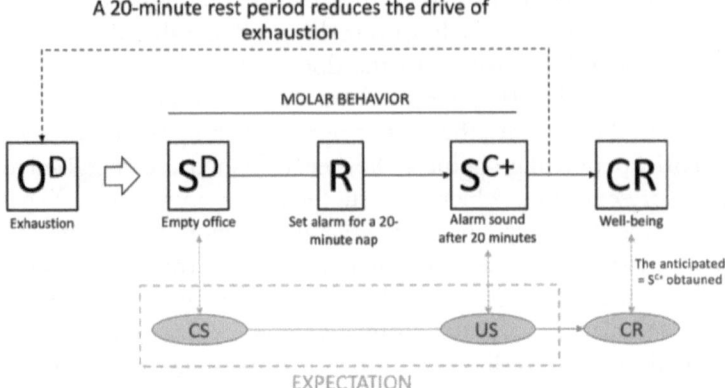

Figure 5.6 Example of functional analysis of digital behavior for alarm drive reduction

In this example, the user sets the cell phone alarm to rest for 20 minutes at work and recover strength for the afternoon presentation. By reducing the exhaustion drive, other drives related to status or dominance are prioritized due to the upcoming afternoon presentation.

behavior (behavioral oscillation), dependent on the presence of competing reinforcers (Hull 1943), although this oscillatory deviation tends to decrease in relation to the strength of the habit (Taylor 1949). In the design of digital behavior, the designer must always be aware of this point. The goal is not to design for 100 percent of users to emit a behavior, but rather to design to increase a desired behavior by a percentage relative to a baseline, which is taken before designing, developing, and implementing changes in the digital service. As designers of digital behaviors, we work to anticipate potential behavioral trends and facilitate their realization for habit formation.

5.5   Segmentation of Users According to Their Drives

Up to this point, it has been explained that when the internal state becomes physiologically or psychologically unbalanced, a metabolic cascade is initiated that triggers internal and external processes to restore the initial equilibrium. These internal processes refer to automatic self-regulation mechanisms of internal homeostasis, such as vasoconstriction and vasodilation to conserve or dissipate heat, or regulate the blood flow to active organs; or the release into the blood of glucose and ketone bodies to fuel internal organs, among others. Conversely, when a behavior is activated, it is because a change in the external environment is required to restore the internal state and preserve both the individual and the species. For example, if despite the internal heat conservation mechanisms, the internal temperature is not restored, a goal-directed behavior to seek an element that provides heat, such as a coat, blanket, or heater, will be initiated. This set of behaviors that are automatically triggered by internal changes is what has been termed instincts. Primary instincts or drives appear to be universal and are originated by the deprivation of some substance or the performance of some behavior for species preservation (e.g., nest-building in birds). Hull enumerated a great number of instinctive or innate behaviors that responded to the need for self-preservation of the organism and generated a primary drive that controlled behavior: hunger, thirst, drowning, pain, cooling, suffocation, constipation, urinary retention, exhaustion (physical and mental fatigue), insomnia, sexual abstinence, and inactivity. Although these instincts seem to be the main ones in humans, and therefore, in the user, as a meta-cognitive and cultural species, other psycho-social drives of origin have also emerged and also need to be solved. William James, in his functionalist approach to psychology, already proclaimed that humans have as many or more instincts than animals, and possibly, these psychological instincts represented a new group of needs (Harlow 1939). Resolving the question of the number of instincts that human beings possess or their origin is not relevant for the purposes of this book. But it is relevant that in modern humans, the internal states that lead them to act have evolved far beyond mere self-preservation. In a way, it could be said that there are secondary drives of meta-cognitive and social origin, which have been historically learned through a process of enculturation and developed insofar as they were associated with primary drives. In this sense, if the drive is activated by an external stimulus whose value is not of innate origin, but learned (money, smiles . . .), we can say that the activated drive is secondary (Baerends 1976; Miller 1951), that is, psychological or representational. Digital services serve to satisfy both primary and secondary drives. For example, the food provided by Uber Eats satisfies the drive for hunger Tinder for reproduction, and Twitter for social contact. In the book "No Filter: The Inside Story of Instagram," it was asserted that "if Facebook is based on friendship and Twitter on opinions, Instagram is based on experiences." While this is an attractive statement for the general public and easy to understand, if we intend to frame it within the context of digital behavior design, it is misstated. Both Facebook, Twitter, and Instagram are digital services to satisfy a psychological need, only that the reinforcers they provide for their satiation differ. While the reinforcers from Facebook and Twitter tend to be more directed to satisfying drives related to communication and social contact, Instagram added to its service new features to manage social reinforcers that would satisfy the drive for status, such as photo filters.

Drives stemming from needs of a psychological rather than biological nature (self-preservation) also constitute a significant source for the emission of goal-directed behaviors or motivated behaviors. Psychological drives emerge in response to social or meta-cognitive needs, and, like biological drives, require periodic satisfaction. Social-type drives are those that fulfill psychological states involving social interactions, while meta-cognitive drives are those that satisfy psychological states concerning self-related thoughts. In the design of digital behaviors, Reiss's model of sixteen basic desires has been adapted as social and meta-cognitive psychological drives (Reiss 2000) and integrated with Shalom Schwartz's types of values (Schwartz 1992), in addition to incorporating others such as identity control (Ryan & Deci 2000) and cognitive validation (Vygotsky 2006). Therefore, psychological drives are divided into two major groups: social and meta-cognitive (see Figure 5.7).

Primary drives, which are self-preservation or biological drives (hereinafter referred to as self-preservation), are divided into two groups: survival and sexual. Self-preservation drives of the survival type are, as previously mentioned, hunger or the need to eat; thirst or the need to drink; pain or the need to escape; cooling or the need to feel warmth; suffocation or the need to feel cool; constipation or the need to defecate; urinary retention or the need to urinate; exhaustion or the need to rest; sleep or the need to sleep; inactivity or the need to move; habitat or the need to have a pleasant shelter; and hedonism or the need to feel pleasure. Sexual drives are reproduction or the need for sexual gratification (including courtship); and offspring maintenance or the need to care for the family. On the other hand, psychological drives can be of a social or meta-cognitive type. These psychological drives may be considered secondary drives, since, in a phylogenetic sense, they can be explained by the dynamics of living in society and innate biological needs. Social-type drives are communicative behaviors or the need to converse with others; social contact or the need for company; honor or the need to obey a moral code (ideology); identity control or the need to manage how others perceive you; status or the need for social importance (includes the desire for attention); and domination or the need to influence (includes leadership and skill development). Meta-cognitive drives are curiosity or the need to explore/know (may include play); revenge or the need to retaliate (includes the desire to regain status or self-esteem); order or the need to organize

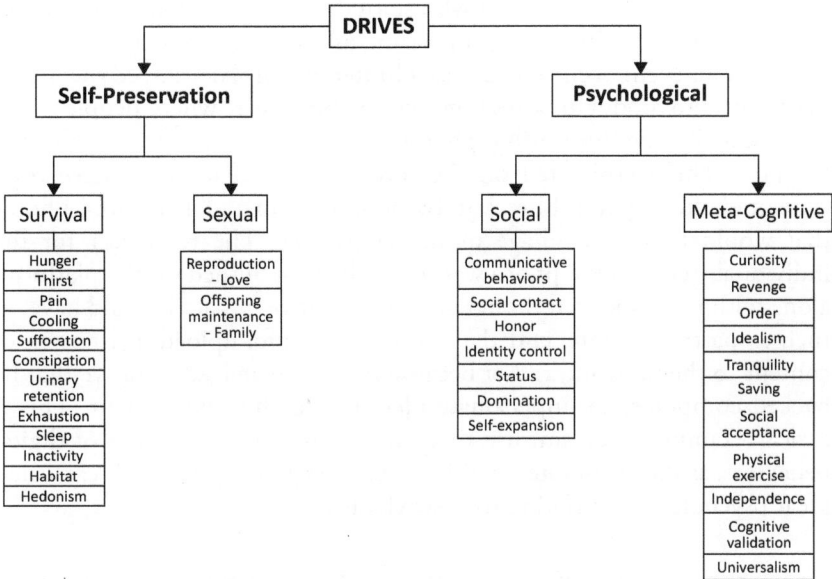

Figure 5.7 Classification of human drives
The following diagram offers a possible classification of the various human drives found in the literature. These have been divided into two broad groups: self-preservation and psychological drives. Self-preservation drives have been associated with individual and species maintenance and are divided into survival and sexual drives. Conversely, psychological drives are related to the subjective experience of well-being, which may be affected by social or metacognitive events, which are related to the individual's own subjective experience of individuality.

(including desires for rituals such as weddings); idealism or the need to improve society (includes altruism and justice); tranquility or the need to avoid anxiety (avoid dangers); saving or the need to collect (valuing economic moderation); social acceptance or the need for approval; physical exercise or the need to exercise muscles; independence or the need to be autonomous; validation or the need for cognitive approval (very common in debates); and universalism or the need for social justice. These are the drives that have been identified for the design of digital behaviors. Surely, other professionals may complement, eliminate, or unify these, or even group them differently. This is at the discretion of those who will use these drives to explain the digital behavior of users.

An example of the misuse of a drive in the design of a digital service was carried out by Hunter Moore, and the drive in question was revenge.

Hunter Moore is a Californian who starred in a Netflix documentary and was dubbed by Rolling Stone magazine as "the most hated man in the world." This event occurred because Hunter, in 2010, developed the digital platform IsAnybodyUp, a revenge pornography site, where people could send sexually compromising photos of their exes as revenge for the breakup. This website fed on the revenge drive that might have been generated after a painful breakup by publishing in a global forum photos that would harm the image of the ex-partner. The reinforcer for the individuals sending the photo was to see them published on the platform, along with the associated thoughts of the negative impact it could have on their ex-partner. Hunter ended up in jail, but not for uploading this type of content to the network, rather because the FBI managed to prove that he hacked computers to impersonate identities for his own benefit. This is a clear example of the misuse that can be given to the design of digital behaviors, as they promote unethical behaviors in people, which can cause great psychological suffering to their victims.

5.6   Emotion As a Source of Energy for Goal-Directed Behavior

Throughout this text, emotion has been ascribed two principal functions: energizing goal-directed behavior and being generated in response to the acquisition of reinforcers. In this section, we will confine our discussion to its energizing function. Emotion is an affective response of an individual to an external or internal stimulus, involving physiological, cognitive, and behavioral changes (LeDoux & Brown 2017). Many of these emotions are characterized by distinctive facial expressions, as well as specific physiological and cognitive alterations (Ekman & Davidson 1994). In the design of digital behavior, emotion is the term assigned to the biochemical milieu that energizes an individual's behavior to resolve the drive, which may be modulated by the anticipation of reinforcer acquisition (see Tables 5.1–5.4).

This emotional state, which energizes your goal-directed behavior, also influences decision-making processes, as the reinforcer obtained will interact with the drive, modifying not only the drive itself but also the underlying emotion (Raghunathan & Pham 1999). Indeed, the role of emotion is to determine the intensity and topography of behavior. This notion aligns with the "feeling is for doing" model, which refers to how emotions can trigger goal-directed behaviors (Zeelenberg *et al.* 2008). This model suggests that emotions provide information about the status of goals and can trigger changes in individuals that facilitate the initiation of behavior.

Table 5.1 *Drives mediating self-preservation of the individual and the species within a biological entity*

Category	Drive	Explanation	Emotion
SURVIVAL	Nutrition	Need for health (Food and Water)	Desperation
	Thirst	Need to drink	Desperation
	Pain	Need to escape	Desperation
	Cooling	Need to feel warmth	Desperation
	Suffocation	Need to feel cold	Desperation
	Constipation	Need to defecate	Desperation
	Urinary retention	Need to urinate	Desperation
	Exhaustion	Need to rest	Desperation
	Sleep	Need to sleep	Desperation
	Inactivity	Need to be active	Desperation
	Habitat	Need for a place to rest	Desperation
	Hedonism	Need for pleasure or self-gratification	Desperation

Table 5.2 *Drives that mediate in the self-preservation of the individual and the species as a biological entity*

Category	Drive	Explanation	Emotion
SEXUAL	Reproduction	Need for sex (including courtship)	Excitement
	Love	Need to feel genuine love	Love
	Offspring maintenance	Need to care for the offspring	Protection

Table 5.3 *Drives that mediate in the psychological state of the individual*

Category	Drive	Explanation	Emotion
SOCIAL	Communicative Behaviors	Need to engage in conversation with others	Loneliness (attention)
	Social Contact	Need for companionship (desire to play)	Boredom, fun
	Honor	Need to adhere to a traditional moral code (ideology)	Loyalty

Table 5.3 *(cont.)*

Category	Drive	Explanation	Emotion
	Identity Control	Need to control others' perception of you	Insecurity, ambition
	Status	Need for social importance (includes desire for attention)	Self-importance, self-esteem
	Domination	Need to influence (includes leadership and skill development)	Efficacy, power, agency
	Self-expansion	Need to enhance rewards through maintaining broad social networks	Relevance, impact

Table 5.4 *Drives that mediate in the psychological state of the individual*

Category	Drive	Explanation	Emotion
META-COGNITIVE	Curiosity	Need to know	Wonder, amusement
	Revenge	Need for payback (includes desire to recover status or self-esteem)	Defense, justice
	Order	Need to organize (including desires for rituals like weddings)	Stability
	Idealism	Need to improve society (includes altruism and justice)	Compassion
	Tranquility	Need to avoid anxiety	Security, relaxation
	Saving	Need to collect (value economic moderation)	Desire to accumulate, own, property
	Social Acceptance	Need for approval	Self-confidence, self-esteem
	Physical Exercise	Need to exercise muscles	Vitality
	Independence	Need to be autonomous	Liberty
	Cognitive Validation	Need for cognitive approval	Self-efficacy
	Universalism	Need for the survival of individuals and the group	Social justice (equality, environmental protection)

Therefore, each emotion aims to energize behavior, in addition to signaling certain characteristics that reinforcers must have to satisfy the drive (Bench & Lench 2013).

There are alternative theories to those suggesting that the drive is the origin of voluntary human behavior, such as the one offered by the American learning theorist Frederick Sheffield (Sheffield *et al.* 1954). Sheffield proposed that it was actually the excitement or emotion produced by consuming the reinforcer or avoiding punishment that prompted reinforcement, not the reduction of the drive. For example, happiness indicates that the reinforcer was successfully obtained, anger signals that the reinforcer was not acquired but may be attempted again; and sadness denotes that a reinforcer was not obtained and there is no hope of future acquisition. The emotional arousal caused by the subjective experience of consuming a reinforcer could significantly influence memory formation (Adolphs *et al.* 2005; Kensinger 2009; Kensinger *et al.* 2007a, 2007b; Mather & Nesmith 2008), meaning that the subsequent emotion induced would act as a unifying element of the sequence of events leading up to the acquisition or loss of the reinforcer. To an extent, the arousal produced by emotion upon consuming/losing the reinforcer can affect the user's concept of the digital service, thereby affecting the likelihood of behavior repetition. Although this proposition was presented as an alternative to Clark Hull's drive reduction theory, in digital behavior it is considered complementary to it. The excitement of obtaining a reinforcer or avoiding punishment modifies the value of the behavior to be performed, thus indirectly it is also selected (Sheffield *et al.* 1951; Sheffield & Roby 1950). That is, behavior selection occurs through the reduction of the drive, but the likelihood of its repetition is determined by the induced emotion. For instance, the loss of value due to delay in reinforcer administration could be explained by the loss of reinforcer properties and the failure to meet expectations, which would generate a negative emotion associated with the performed behavior, and thus, with the utilized digital service. Consider the following case. Alice is very hungry and decides to order a delicious pizza from her favorite restaurant, "El Veneciano" through Uber Eats. Her expectation of the reinforcer (the pizza) is that it will arrive in twenty-five minutes (as indicated by the app) and be tasty. Due to various reasons, the pizza takes fifty minutes to arrive, and to make matters worse, it is cold upon arrival. The fact that the pizza was delayed more than expected and was not as tasty as anticipated, causes an emotional reaction of anger in Alice (expectation differs from reality), which affects the likelihood of repeating the same behavior when she wants to eat "El Veneciano's" pizza again. Although emotion can affect the probability of engaging in the behavior again, it should not be analyzed in isolation. In the next scenario when Alice considers ordering a pizza from "El Veneciano" through Uber Eats, she will consider within her cognitive unit various goal-value representations such as her history of obtaining reinforcers from "El Veneciano" via Uber Eats, environmental pressure (does she have time to go to the restaurant to pick it up?), whether she likes pizza from other restaurants, whether she would mind eating a different type of food, etc. (FitzGibbon *et al.* 2020). The variability or fluctuation in response is the consequence of a multitude of external and internal factors affecting decision-making. Therefore, for the design of digital behavior, reinforcement is a process that ideally occurs when the drive is reduced and a positive emotion is obtained.

CHAPTER 6

Human Beings As Psychological Entities

6.1 Social Interaction As the Driving Force of Goal-Directed Behavior

Evolutionary theories suggest that instinct is the source of the complex and intelligent behaviors currently observed in social environments.[1] The phylogenetic history of our species has seemingly favored genetic profiles with psychological traits predisposed to human social interaction. Social interaction is a fundamental element in both the phylogenetic and ontogenetic evolution of humans, as discussed in previous chapters. Such interactions have been crucial in the development of our brains, which have evolved to double the size of those of our closest relatives, chimpanzees (Rilling 2014). Our brains are equipped with a "genetic starter kit" of basic behavioral programs that favor certain activities over others, thus facilitating an organism's adaptation to its environment (Heyes 2014). In humans, this "genetic starter kit" comprises a set of rules that manifest as instinctive or innate behavior, activating automatic motor and cognitive patterns that trigger certain survival-favoring behaviors and integrate humans into communal living. The phylogenetic history of humans, and therefore their historical evolution, should be studied with the understanding that it has occurred under significant environmental pressure dominated by a social and cultural environment. Consequently, natural selection has been favoring the emergence of new prosocial instincts that are genetically transmitted (Enfield & Levinson 2020). This phylogenetic selection of social instincts has endowed humans with basic cognitive mechanisms for cultural learning, which allow us to acquire the tools needed for survival and reproduction within a specific

---

[1] In the scientific and psychological realm, some experts prefer the use of terms such as "fixed action pattern" and "modal action pattern" over the concept of "instinct." This preference is justified due to the complexity involved in distinguishing between elements of behavior acquired by an individual through environmental experience (learned) and those that are the result of genetic information (nonlearned) (Baerends 1976; Barlow 1968).

geographical and cultural setting. Among these socio-cognitive instincts we inherit is a particularly special one, "social motivation." Compared to other nonhuman primates, the pursuit of social contact holds significant reinforcing value for humans, which seems to be genetically preordained. For instance, humans are predisposed to detect the movement of other animals, including humans, more rapidly than that of non-animal objects or those that are inanimate (New *et al.* 2007). In one experiment, individuals were exposed to images containing other people, animals, and moving cars. It was found that people could detect moving people and animals with over a second's advance notice compared to cars. The results of these experiments have been explained through the animate monitoring hypothesis (Calvillo & Hawkins 2016), which suggests that the attentional mechanisms triggered by the automatic monitoring of animal activity are set off by inherent characteristics of the beings, rather than inherent characteristics of the movement. Additionally, there is a tendency for human infants to engage and sustain joint activities with other humans, whereas in chimpanzees, despite some intentionality, it is not as marked as in human babies (Tomasello & Carpenter 2005; Warneken & Tomasello 2006). Regarding this, it has been reported that the perception of social status is present in both primate and nonprimate children from an early age, showing a preference for interacting with peers of favorable social status (Herrmann *et al.* 2013). The reinforcing effect of social activity in humans can be explained by the release of a "social biomolecule" known as oxytocin, which can suppress appetites (Lawson *et al.* 2015) and anxiety (Churchland & Winkielman 2012). That is, social contact releases oxytocin, and this oxytocin can change the internal state of the organism by reducing drives such as hunger and stress (Churchland & Winkielman 2012). Intriguingly, this peptide is found in all beings, including insects and worms (Gruber 2014), indicating that interaction with peers may be present to some degree throughout the animal kingdom. Having high "social motivation" are that it allows the brain to show a preference for social environments over isolation (Apicella & Silk 2019). Therefore, there seems to be an innate human predisposition to seek out such social reinforcers as smiles, caresses, kind words, nods ... in a way that strengthens our group ties with other members.

6.2   Social Behavior Emerges within the Cultural Context

Social instincts, seeking satiation through social interaction, have evolved as an alternative pathway to satisfy more fundamental stimuli (Declerck &

Boone 2018; Hilbe *et al.* 2018). As eloquently expressed by Allport, "the pursuit of literature, the development of good taste in clothes, the use of cosmetics, the acquiring of an automobile, strolls in the public park, or a winter in Miami – all may first serve, let us say, the interests of sex" (Allport 1937). This controversial statement aims to explain that certain drives can emerge from the need to obtain more basic drives. According to Vygotsky, the predominant instincts in an animal and a child follow distinct ontogenetic developments due to cultural mediation (Vygotsky 1978). These primary physiological needs in children are transformed into secondary, socially originated needs related to how the instinct should be resolved. Lewin referred to these new social-type needs as quasi-needs from which a genuine concern for the outcomes of the social resolution of the instinct emerged, indicating access to the satiating stimuli (Lewin 1926). It could be suggested that our instinctive behavior has evolved in parallel with culture. Like other animals, humans engage in exploratory behaviors (appetitive phase) when we need to satiate a drive such as hunger. When we are hungry, we embark on an exploratory search for signals (discriminative stimuli) that indicate the presence of food. Intriguingly, this discriminative stimulus signifies that an individual must initiate a culturally learned and mediated behavior to obtain the food reinforcer. For instance, the value of a sign bearing the McDonald's logo is directly proportional to a person's hunger and their access to alternative food sources.[2] This sign indicates that by initiating a culturally mediated behavior, one can obtain food. This mediated behavior is the exchange of money, a stimulus with a socially derived value. This appetitive phase has evolved with culture and is subject to modifications to adapt to the new social and technological environment of the moment[3] – a technologically mediated adaptation, for example, could be the use of food delivery apps to avoid going to the supermarket or restaurant.

---

[2] In a dystopian world, McDonald's could implement a strategy that promotes a constant sensation of hunger in consumers, aiming to maximize their revenues.

[3] For Lorenz and Tinbergen, the term "instinct" did not necessarily exclude the presence of learning processes in the ontogeny of instinctive behavior in species. Lorenz discussed "instinct conditioning intercalation patterns" (Instinkt-Dressurverschrankungen), and particularly emphasized the role that "imprinting" could play in such behavioral constructs. Furthermore, Tinbergen provided examples of appetitive behaviors that were controlled by instinct and involved the incorporation of learned components at various levels of the instinct's hierarchical organization. The term "modal action pattern," previously mentioned, refers to the fact that these intercalated patterns were not found; instead, learned and unlearned behavioral patterns are intermingled. Hence, the preference for using this term instead of instincts by several psychologists and ethologists (Baerends 1976).

Three historical waves can be delineated in which the appetitive phase for the satiation of an instinct has changed within human culture. The first wave spans from the Paleolithic to the Neolithic period, where humans, as hunters and gatherers, had to acquire the satiating stimuli directly from nature without cultural mediation. When someone was hungry, they had to hunt. The second wave can be situated around the Neolithic period, with the emergence of the first settlements and the development of information and communication technologies, specifically the telephone. During this period, regardless of distribution sophistication, the reinforcers satisfying instinctive behaviors were mostly administered culturally. Although since the Neolithic era to the present day, there still survive groups of individuals who go out to obtain their reinforcers (e.g., hunters), for the vast majority, the acquisition of reinforcers is mediated by a cultural structure. For example, to obtain food we go to a supermarket or restaurant, while to seek a sexual partner, we tend to visit leisure venues such as bars or nightclubs. At this juncture, instinctive behavior is strongly governed by culture. Finally, the third wave, in which we currently find ourselves, originated from the advent and development of information and communication technologies. For the first time, digital behavior appears in full force, as it is not necessary to go to a specific physical location to obtain your reinforcer; instead, one can emit a digital behavior on a technological tool to obtain the reinforcers. While many services could be accessed through the conventional telephone, it was not until the implementation of communication protocols in portable phones in the first decade of the twenty-first century that online digital services became globally widespread. Furthermore, these online digital services exploded and occupied a significant part of business due to the coronavirus pandemic at the end of 2019. The pandemic led to the global population being confined to their homes and using these digital services massively to satisfy our drives. Whereas in the first wave, the acquisition of the satiating stimuli for instinctive behaviors is direct, in the second and third waves, the acquisition of these reinforcers is indirect. During the second wave, the mediation is cultural and in-person, that is, the individual has to move to a place or establishment to receive the satiating stimuli. In contrast, in the third wave, the mediation is technological and online. There is no need for personal displacement, as the desired items are delivered to one's residence.

6.3   Attaining Status As a Primary Social Drive

Status can be regarded as a secondary, social type of drive. Social groups are organized into hierarchies, where members hold varying levels of power, influence, skill, or dominance (Fiske 2010; Magee & Galinsky 2008; Mazur 1985; Zitek & Tiedens 2012).[4] Humans have a propensity to view the world through a social lens, discerning the best, brightest, or most privileged within our social group. Moreover, this hierarchical structure is also evident in other animal species (Anderson *et al.* 2001; Berger *et al.* 2003; Magee & Galinsky 2008), suggesting that status may have a common evolutionary origin (Mazur 2013). Group members can rapidly identify the social status of a peer based on various stimuli or events that typically accompany such descriptions, including job position, level of education, car brand, attire, or neighborhood of residence. Typically, members with higher status within a social group are perceived to possess greater power, influence, and other privileges compared to those with lower status (Fragale *et al.* 2011a; Mazur 2013; Zitek & Tiedens 2012). Those with higher status can access more and superior resources and benefits, significantly impacting their quality of life. The purpose of status is to organize the social group for the distribution of reinforcers such as potential mates and food (Sapolsky 2005), facilitate social learning (Henrich & McElreath 2003), and enhance individual motivated behaviors (Halevy *et al.* 2011a; Magee & Galinsky 2008). The sensitivity to status is a cognitive phenomenon that can emerge early in development. For instance, it has been observed that both children and adolescents are particularly sensitive to information that reveals a person's status. In adolescence, the fear of rejection is a significant force in motivating and guiding social interactions (Asher & Cole 1990). Furthermore, among adolescent groups, status is what primarily differentiates their members, more so than other individual achievements (LaFontana & Cillessen 2010). This particular perception of status during adolescence is brought forth by the onset of cognitive abilities that recognize the rewards of interacting with others; in addition to the capacity to project the long-term consequences so as to adjust goal-directed behavior to avoid status loss (Asher & Cole 1990; Koski *et al.* 2015). Conversely, in childhood, the indicators of an individual's social status seem to be different. A series of experiments showed that, in children aged 9 to 12, being humorous correlated with likability and popularity. Children showed a preference for peers with whom they had fun. In this sense, fun

---

[4] The terms "status" and "rank" are often used interchangeably, as both represent the positions that members of a group occupy within a social hierarchy, usually designated by a numerical value within an ordinal ranking (Chiao *et al.* 2004, 2009). It is also important to distinguish between the concepts of power (which reflect the control over resources and the ability to influence others through reinforcers and punishments) and status (the capacity to influence the group due to admiration and respect from other members) (Blader & Chen 2014; Magee & Galinsky 2008).

may be one of the greatest social reinforcers a child can experience (Laursen *et al.* 2020). These findings are intriguing when designing virtual worlds for children. This demographic group will primarily seek reinforcers that guarantee enjoyment, in the sense of recreational activities with other children as well as the attainment of objects that allow for increased amusement and the performance of enjoyable activities. Platforms such as Roblox have successfully identified these hedonic needs of children and have leveraged them to create a virtual platform where children can create their own games and share them with others.

Status can be measured through social opinion or reputation, and is generally associated with admiration and respect (Anderson *et al.* 2006; Fragale *et al.* 2011b; Gould & Bearman 2002). The ease with which status-indicating stimuli are perceived, and the readiness with which they are ranked within the social hierarchy, reflects a preference for social organization. This predilection for social structuring is likely driven by a desire to know where we stand in relation to others to define our social role and thereby promote social interactions relevant to the individual. The higher a person's position in the social structure, the more relevant the interaction with them, and the more access to reinforcers they will have (Halevy *et al.* 2011b; Savin-Williams 1979). Indeed, having status within a community allows one to access certain reinforcers or even to enhance the value of the reinforcers one can already access such as sexual partners, material goods like houses, cars, food ... and even to new leisure activities that have a greater emotional impact (e.g., better seats at sports stadiums). Therefore, the regulation of status becomes one of the main drives of social beings, especially in humans. Since status depends on the signs and signals that accompany a person when interacting with other group members, controlling these signs becomes key to status regulation. As Adam Smith noted in his book "The Wealth of Nations": "With the greater part of rich people, the chief enjoyment of riches consists in the parade of riches, which in their eye is never so complete as when they appear to possess those decisive marks of opulence which nobody can possess but themselves." Thus, regulating our status is not a new and modern phenomenon, but rather, throughout the history of human civilization, it has acted as a drive that explains many of the behaviors that have occurred. What is new, thanks to social media, is the display of our status twenty-four hours a day on our digital profiles. For the design of digital behavior on social networks, the regulation of status is achieved through controlling one's own identity.

6.4 Identity Control: The Emerging Drive from Social Media Networks

Identity is a psychological construct that delineates an individual's self-perception within a social group. The identity users align with can modify their digital behavior (Ashokkumar *et al.* 2020). In essence, it is the interaction among group members in diverse social situations (Mead 1934), mediated by symbolic exchange,[5] which transforms an individual's identity and hierarchy, thereby also reconfiguring the social organization of the group (Burke & Stets 1999). In this regard, the drive for identity control is a focal theme within social networks. Edward Deci and Richard Ryan have previously identified identity control as an innate need to feel self-regulated and to possess a greater sense of self-determination in one's life (Ryan & Deci 2000). William James, the father of American psychology, posits that humans do not have a single identity but rather adapt it across various contexts, roles, or groups (Burke & Stets 1999). A unified perspective on how identities shift may be found in Charles Cooley's "looking glass self theory" (Colley 1902). The design of digital behaviors draws upon this theory to elucidate how individuals employ social networks to control their identity. Cooley suggests our identity is forged from the perceptions we believe others hold of us. For instance, if I perceive that many people regard me as very sociable, I will consider sociability a facet of my identity. Conversely, if I believe that others view me as worthless, then being ineffectual becomes part of my identity. Thus, altering my identity necessitates influencing others to change their opinions of me. One rationale for the compulsive use of social networks is justified by the intent to self-shape identity through the creation of content, such as comments, videos, or photos, that can alter how others perceive us. Or, under optimal circumstances, to ensure that those consuming the content shift their perspective about the individual, thereby forming a more favorable opinion. Features like "likes" or comments are nothing more than a system of administering social reinforcers, allowing the content creator to garner group validation for their identity (self-meaning).

---

[5] The implementation of symbolic exchanges is a crucial part of the planning for digital behaviors. In virtual reality environments, these interactions may take the form of bodily, facial, linguistic expressions, gestures, and dances (known as emotones), which are produced via avatars. Conversely, in digital services where avatars are uncommon, the use of language or other means of symbolic communication such as emojis.

However, this represents a double-edged sword, as content is uploaded to social networks with the expectation of achieving a certain number of "likes" or positive comments. Yet, often the opposite occurs – either the anticipated number of "likes" is not met, the desired individual's "like" is not obtained, or unpleasant comments are received, causing harm to one's identity within the group. Frequently, such aversive stimuli may lead the user to respond defensively or to delete the content due to social devaluation, as well as the user's desire to restore their prior social status or mitigate damage (Robertson *et al.* 2018).[6] If these negative comments or the absence of anticipated social reinforcers decrease the likelihood of future content posting, this could be considered a form of punishment conditioning. Corporations behind social networks are aware of this phenomenon and have devised sophisticated gratification systems to secure social reinforcers and control our identity. Beyond traditional reinforcers such as comments and "likes," other social reinforcers have emerged throughout the history of social networks or other digital services incorporating social profiles. A prime example is the tweet analytics provided by Twitter or Spotify's end-of-year "Wrapped" feature. "Wrapped" is a digital artifact reflecting our musical preferences, which can be shared on other networks to gain social reinforcers that positively influence our identity. They may also satisfy users' curiosity drive by revealing more about their tastes, but they are primarily designed for sharing. It is vital to acknowledge this new reality where the concept of identity has evolved with the advent of new technologies. Identity control is a potent drive for predicting user behavior, not only in "real" environments but also in digital and virtual ones. Although discussed here alongside status, identity control can be considered a distinct drive from status, as it involves regulating identity not only to manage status but also for subjective personal well-being (Sumner *et al.* 2014).

---

[6] Here, it is important to understand the emotions induced by those aversive stimuli to know how the "attacked" user might act. If it induced shame, they might offer explanations, while if it induced anger, they might interact more aggressively. Or they might even cease to interact if they perceive the environment as hostile (Edelmann 2013).

CHAPTER 7

Behavioral Competition

7.1 Behaviors Compete for a Limited Amount of Time

Any behavior, before being enacted, exists in a potential state that will collapse into a specific molar behavior when the subject decides to initiate the sequence of interactions necessary to satisfy a specific need. This jungle of behavioral probabilities collapses into a single behavior that blocks the remaining possible behaviors and determines the subject's fate. There are certain conditions that will limit this "battle" for survival occurring among all the potential behaviors to be enacted. The first of these could be defined as the arrow of time, which implies a change in the initial conditions with which the behavior is performed. For instance, the passage of time, along with the consequences of our own actions, make it impossible in a situation of regret to undo what we have done and select another behavior to obtain different consequences. In life, there is no "save game" feature for us to retry a task in the same conditions if we fail. This constraint could be defined as the "entropic limit,"[1] and there is little we can do about it other than accept and take responsibility for the consequences of our actions. On the other hand, there is another limitation more related to cultural aspects and how the activities we perform are distributed over time, known as "behavioral competition." Days have a finite number of hours, and usually, at each hour, minute, and second we are behaving, that is, performing some action (see Figure 7.1).

---

[1] The Arrow of Time, a postulate formulated by the distinguished British philosopher and physicist Arthur Stanley Eddington, addresses the direction and irreversible nature of time in relation to the second law of thermodynamics, which posits that the entropy of the universe increases steadily. Although this is a controversial concept that raises fundamental questions regarding the essence of time, the Arrow of Time is essential for understanding the irreversibility of events and how variations in energetic states (entropy) affect different systems, including human beings and the natural environment. This proposition suggests that time moves in a specific direction, facilitating a series of irreversible events that continuously evolve as the entropy in our universe and the world around us grows (Ben-Naim 2020; Popper 1965).

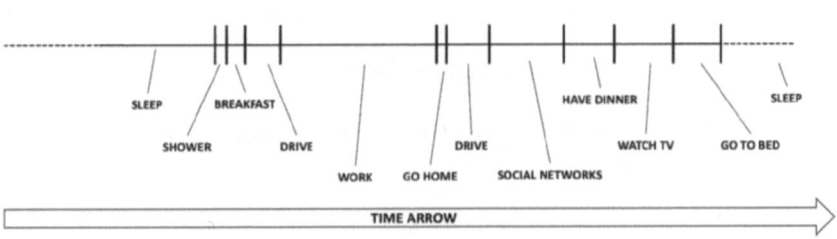

Figure 7.1 Behavioral pattern distributed over time
The illustration depicts how an individual might distribute their behavioral repertoire
throughout an ordinary day. These behaviors occupy a specific amount of time and are
aimed at fulfilling or satisfying certain short-term and long-term goals. The behaviors
performed can be of two types: self-preservation or psychological. The study of behavioral
patterns and the clustering of users who exhibit similar behavioral patterns is known as
"behavioral profiling" (Canhoto & Backhouse 2008). Behavioral profiling involves
recording the events and actions of users to determine a typical behavioral model and to
study the behavioral deviations from their typical model.

The digital behavior designer must understand that people, and therefore users, are creatures of habit. Normally, users occupy their twenty-four hours a day with activities that provide them with certain reinforcers necessary for the satisfaction of self-preservative and psychological needs. Therefore, any new behavior that one wishes to establish must be positioned within a time space already assigned to an activity, where it will compete with other behaviors for the user's time. Indeed the competition between behaviors is to occupy the temporal space of other behaviors. Let's imagine that we are the authors of a psychology podcast, and we want users to listen to our podcast. As an initial strategy, there are several possibilities: encourage users who already listen to psychology podcasts to listen to yours, encourage people interested in psychology who listen to podcasts to listen to your psychology podcast, incentivize people who do not listen to podcasts but are interested in psychology to listen to your podcast ... there are many possible combinations. Suppose the podcast author decides to start the campaign for people who already listen to psychology podcasts. How would they get these people to listen to their podcast? The first step is to be aware that these people dedicate some time of their day or week to listening to podcasts of their interest. It may be that they listen while doing sports, commuting to work, while at work, while washing dishes ... During this time, the person dedicates their time to listening to their favorite podcasts, so the new podcast would enter into competition with these, stealing minutes or even suppressing their listening. As seen in previous chapters, the behavior of listening to these podcasts is controlled by the reinforcing stimuli that the subject receives from listening, which can be direct (e.g., podcast content) or indirect (e.g., eliminating the feeling of tiredness if used while doing sports). Here the podcast author would need to study their market well, know what needs they want to satisfy, provide the appropriate reinforcers, and eliminate potential punishments. Providing authors with tools that improve the content of their podcasts both technically (noise filters, etc.) and narratively (artificial intelligence algorithms that highlight unnecessary content, etc.) can facilitate users preferring it over the competition. This rapid analysis can also be useful for digital behavior designers to truly understand that the digital service they are designing competes with other behaviors that are already occupying a space of the user's time.

7.2 Contingent Reinforcer Matrix for a Single Behavior

According to the US Bureau of Labor Statistics, in 2021, the allocation of time Americans spent on primary activities was as follows: sleeping (8.96 hours), eating and drinking (1.22 hours), household chores (2.47 hours), shopping for goods and services (1.68 hours), caring for and assisting family members (2.17 hours), working (8.08 hours), educational activities (5.39 hours), social and religious activities (2.00 hours), and leisure and sports (5.43 hours) among others (US Bureau of Labor Statistics 2022). Although this distribution of time and behaviors can vary based on each individual's learning history, the time dedicated to each behavior will depend on the reinforcers received for performing it. Let us consider a hypothetical case involving the use of digital services such as social media. Imagine that George, an average American, has developed a habit where he uses social media before going to sleep (see Table 7.1). He dedicates one hour to this activity before he stops and goes to sleep.[2]

Table 7.1 presents the possible reinforcers and punishments that each social network could administer, which would vary according to the user utilizing them. Up to the acquisition of this habit, these social networks would have "employed" their reinforcers and punishment, competing among themselves to occupy as much of George's time as possible that he dedicates to looking at social media before going to bed. As the

---

[2] These studies suggest that excessive use of smartphones and exposure to their screens at bedtime are associated with poor sleep quality and other potential disturbances (Alshobaili & AlYousefi 2019; Randjelović *et al.* 2018).

Table 7.1 *Contingent reinforcer matrix for using social networks while lying down in bed*

		USE OF SOCIAL NETWORK				
S^D	R	S^C Description	Type Consequences	S^C Contingency	S^C Contiguity	S^C Mediation
In bed lying down	Use of social networks (Facebook)	Finding friends' posts	Reinforcer	Positive	Immediate	Direct
		News of interest	Reinforcer	Positive	Immediate	Direct
		Statistics of my economic activity	Reinforcer	Positive	Immediate	Direct
		Interactions in an online game	Reinforcer	Positive	Immediate	Direct
		Long videos	Punishment	Positive	Immediate	Direct
		Interactions with family members	Punishment	Negative	Deferred	Indirect
		Information overload – not selective	Punishment	Positive	Immediate	Direct
		Few interactions (few friends)	Punishment	Negative	Immediate	Direct
S^D	R	S^C Description	Type Consequences	S^C Contingency	S^C Contiguity	S^C Mediation
In bed lying down	Use of social networks (Instagram)	Funny videos	Reinforcer	Positive	Immediate	Direct
		Informative videos from influencers	Reinforcer	Positive	Immediate	Direct
		Video production with filters and other edits	Reinforcer	Positive	Immediate	Direct
		Low cognitive effort – few words	Reinforcer	Negative	Immediate	Direct
		Self-verbalizations indicating the need to post content	Punishment	Positive	Immediate	Direct

S^D	R	S^C Description	Type Consequences	S^C Contingency	S^C Contiguity	S^C Mediation
In bed lying down	Use of social networks (Tik Tok)	Funny videos	Reinforcer	Positive	Immediate	Direct
		Informative videos from influencers	Reinforcer	Positive	Immediate	Direct
		Direct	Direct	Direct	Direct	Direct
		Low cognitive effort – few words	Reinforcer	Negative	Immediate	Direct
		Short videos – more variety	Reinforcer	Positive	Immediate	Direct
		More personalized content – algorithm	Reinforcer	Positive	Immediate	Direct
		Anonymity	Reinforcer	Negative	Immediate	Direct
		Can become viral faster (more interactions)	Reinforcer	Positive	Delayed	Indirect
		Not serious	Punishment	Negative	Immediate	Direct

competition between different networks unfolds, they settle into a certain usage time, until George stably allocates his time among them, reaching a steady state. In this case, let us suppose that the steady state was set with forty minutes on TikTok, fifteen minutes on Instagram, and five minutes on Facebook. This time distribution seems to assign a value to each social network for George. Staddon notes that in free conditions, where an organism is given free access to different activities, it will dedicate a certain amount of time to each of them, and that percentage of time will reflect the probability of engaging in that activity again and, consequently, the degree of preference for that activity (Staddon 2003). According to this author, modern theories of utility accept that behaviors or events can be rank-ordered in terms of value (ordinal scale), although magnitude cannot properly be assigned to utilities (satisfaction) (Staddon 2003).

Modern economic theory of revealed preferences suggests that the best way to measure customer preference is by observing their consumption behavior of products or services, which will be affected by internal variables of the individual such as temporal preferences, risk preferences, social preferences, and cognitive limitations in decision-making. That is, the user is an imperfect machine of behavior (irrational decision-making), which would cause their consumption behavior not to always be stable (Tipoe *et al.* 2022). In the previous example of how a user distributes their time before sleeping for the consumption of social networks, the satisfaction obtained by the user (a concept called "utility" in economic theory of individual consumption behavior) will determine their preference for the different social networks (Stigler 1950). These economic principles use the concept of utility obtained by the user to explain decision-making in the consumption of certain assets, with the "law of diminishing marginal utility" and the "utility maximization model" being two proposals to explain consumer behavior. The law of diminishing marginal utility indicates that the amount of utility of a product/service decreases as the amount of use increases (Ormaxabal 2006). This is a clear example of the satiation effect of the reinforcer explained by operant conditioning; the more access one has to the reinforcer, the more its value decreases. In the previous example, the satisfaction produced by TikTok's reinforcers decreases as the user obtains them continuously (hence the importance of introducing intermittent reinforcement schedules). That is, the satisfaction produced by TikTok's reinforcers when you first open the app is not the same as when you have been using it for thirty minutes. On the other hand, the utility maximization model makes predictions about how a consumer will spend their money or resources to obtain the maximum possible satisfaction when consuming the product (Board 2009). In the previous example of social networks, this model could be used to predict users' decision-making when distributing their time in the consumption of products/services to maximize satisfaction. That is, if George uses TikTok for forty minutes, it is because that is the time he has deemed appropriate to maximize his satisfaction. In the design of digital behaviors, the satisfaction obtained by the user is a key factor in anticipating their behavior.

The satisfaction of a user when using a digital tool is determined by how the reinforcer has been obtained and the drive with which the user came to consume the digital service has been satiated. The economic concept of utility is an emotional response resulting from the satiation of a drive and the hedonic properties associated with the consumption of the reinforcer. This resulting hedonic satisfaction from the use of technological services and products is a variable to consider in the design of digital behaviors, as it can often even cover goal-directed behavior for the satiation of a drive. In this sense, the release of dopamine in the brain can overshadow survival drives, stopping the search for satiating reinforcers in favor of these new stimuli that activate the release of dopamine. In the 1950s, in an experiment by James Olds and Peter Milner, rats were placed in a specific location within a large box to administer electrical stimulation to their brain's pleasure centers (septal area, mamillothalamic tract, and anterior cingulate cortex) (Olds & Milner 1954). This brain stimulation caused the rats to associate the location in the box with pleasure, so they returned to this place when placed back in the box. This suggests that it could be the release of dopamine and the consequent emotion from consuming the reinforcer that reinforced the behavior. This hypothesis had already been stated by Frederick Sheffield, through the Theory of Induced Emotion. Subsequently, various researchers inquired into what would occur if control over brain stimulation was yielded to rats. The findings were astonishing, as the rats displayed a preference for self-administration (by pressing a lever) of an electric shock in the pleasure centers over eating or drinking when they were hungry and thirsty (Frank *et al.* 1981; Rossi & Stutz 1978), even to the point of risking death by inaction to feed (Routtenberg & Lindy 1965). In some manner, the pleasure induced by dopamine release could overshadow or mask the aversive sensation caused by hunger, thus the rat did not feel the physiological urgency to feed or drink. These results could not be replicated if the brain stimulation occurred in other cerebral regions apart from these pleasure centers. In this regard, it has been demonstrated that when electrical stimulation is carried out in areas responsible for the release of dopamine (ventral tegmental area), the self-deprivation of food and water depends on whether the reinforcing value of the shocks is greater than the value of food and water, even in a state of need (Frank *et al.* 1984). It appears that there might be indications that the acquisition of pleasure is prioritized over the suppression of the drive. Therefore, it is emotion that energizes goal-directed behaviors, not the drive itself.

A paradoxical example concerning this is the satiation of the drive of curiosity, which can sometimes even overshadow aversive experiences.[3] Curiosity can predict risky behaviors such as initiating tobacco use, exposure to electroshocks, and voluntary exposure to information that may have a negative effect, such as images of mutilation or blood (Kruger & Evans 2009; Oosterwijk 2017). In an experiment, participants were asked whether they preferred to view an image of a mutilated finger or have it described to them verbally. The majority chose to view it (Oosterwijk 2017). To understand this phenomenon, it has been suggested that individuals may also seek information in a noninstrumental manner, that is, not valuable for the future (for instance, information about celebrity gossip) (Gottlieb & Oudeyer 2018). In fact, some studies demonstrate how humans (and even some animals) would incur a significant energy cost just to receive noninstrumental information (Gottlieb & Oudeyer 2018; Rodriguez Cabrero *et al.* 2019; Wang *et al.* 2019). Furthermore, it seems that there is beginning to be a consensus among different fields of psychology and neuroscience that obtaining information without instrumental value can be a drive strong enough to block the attainment of reinforcers such as food or water (Gruber & Ranganath 2019; Rodriguez Cabrero *et al.* 2019). Some authors additionally suggest the existence of an "incentive salience," which can be a feeling of pleasure from "waiting" for a reinforcer, distinct from the hedonic component of "cognitive desire," or the anticipation of the emotional benefits of consuming it (Berridge 2009; Berridge *et al.* 2010). The enjoyment produced by the anticipation of the reinforcer and also by the anticipation of its hedonic characteristics might explain some of the efforts individuals make without the certainty of obtaining an appetitive stimulus (Robinson & Berridge 2008). Thus, this could also account for certain behavioral fluctuations that occur in the individual behavior of users. Hence, the intrinsic hedonic pleasures of stimuli that may emerge

---

[3] The sensation of curiosity may give rise to an "aha-moment," which has been linked to dopaminergic structures, also part of the pleasure centers (Tik *et al.* 2018). Therefore, the feeling of curiosity may also be governed by the experience of pleasure.

during digital behavior, and the reinforcing action of the behavior generated by curiosity or by the anticipation of a potential reinforcer, should also be considered in the design of digital services.

7.3 Behavioral Contrast As an Explanatory Theory of Competing Behaviors

In psychology, various theories have emerged to predict the amount of time a person will dedicate to a particular activity. These include additivity theory or by differential contingencies associated with reinforcement schedules (Gamzu & Schwartz 1973); competition theory or the change in the value of a reinforcer (Ettinger & Staddon 1982; Hinson & Staddon 1978); matching theory or the proportionality between the response rate and the quantity/duration of the reinforcer (Herrnstein 1970) and habituation theory or the prevention of habituation of the reinforcer to increase its value and effectiveness (McSweeney & Weatherly 1998). While none of these theories has yet prevailed over the others, they all rest on the premise that behaviors compete for reinforcers based on the value individuals assign to them. For instance, if George sleeps with his partner who is annoyed by the sound of TikTok and Instagram videos, a shift in the value of reinforcers occurs, prompting George to spend more time reading X (formerly Twitter) tweets and less time watching TikTok videos. To better elucidate this shift in reinforcer values and the ensuing competition among behaviors, let us assume that George reaches a steady-state distribution of social media use before sleep, predominantly using TikTok because his favorite influencer mainly posts humorous videos on this platform. For George, these humor videos serve as his primary reinforcer, as they relax and distract him (negative reinforcement) after a particularly stressful day. Suddenly, one day, he discovers that his favorite influencer has stopped posting on TikTok and started on Instagram.[4] At this juncture, the steady-state distribution of social media usage is disrupted due to a change in the value of the reinforcer associated with Instagram use, which then begins to vie with TikTok use for more minutes in George's pre-sleep social media routine. "Behavioral contrast" (Reynolds 1961) is a concept that accounts for the competition among latent behaviors as a consequence of the introduction of a new reinforcing stimulus or a shift in the value of existing reinforcers (evaluated in terms of frequency, precision, intensity, duration,

---

[4] The endorsement of certain influencers to broadcast content exclusively on a specific digital platform has become a common practice.

etc.). The time lost by one behavior is gained by another, a phenomenon economically known as a "zero-sum game." For example, if George allocates forty minutes to social media before sleep and begins to use Instagram more, this will result in decreased TikTok usage so that after redistributing the time dedicated to each platform, the total remains at forty minutes. This exemplifies a "zero-sum game," which is precipitated by a behavioral contrast stemming from a change in the value of the reinforcers for one of the competing behaviors (e.g., favorite influencer now on Instagram). Another real-life instance occurred on October 4, 2021, when Meta Corporation (owner of Facebook, Instagram, and WhatsApp) experienced a server outage lasting over six hours (The New Statesman 2021). WhatsApp was literally nonfunctional, thus ceasing to dispense reinforcers, which instigated a behavioral competition with other latent behaviors such as using the messaging service Telegram. This incident appears to have significantly benefited Telegram, which acquired millions of new users. When service was restored, the majority of users returned to WhatsApp, not merely because it was operational again, but because the reinforcers it provided were once again available. Nonetheless, many users remained with Telegram or began using it sporadically. The Facebook outage also led to competition with other latent behaviors, like increased searches for information through alternative channels (NiemanLab 2018). Indeed, behaviors compete to fill a space in the user's limited time, and many require digital technologies to manifest and secure the appropriate reinforcers to reduce drives. Behaviors are latent because drives are always present. This point is crucial because, in the design of digital behaviors, it is not the features of the digital service or product that take center stage, but the reinforcers they provide if utilized. Hence, for the digital behavior designer, the digital service does not compete with other digital services; rather, it is the reinforcers they offer that are in contention. For this reason, it is imperative to be constantly designing new reinforcers or enhancing their value to solidify the use of the digital service or to enter into competition with other digital services (see Figure 7.2).

7.4 Functional Characteristics of Behavioral Contrast

In behavioral competition, two theoretical elements must be considered to understand the direction of behavioral change: the changed component versus the unchanged component (Hinson & Staddon 1978). If behavioral change occurs because the reinforcement of one behavior is altered in some parameter that negatively modifies its value (changed component), the

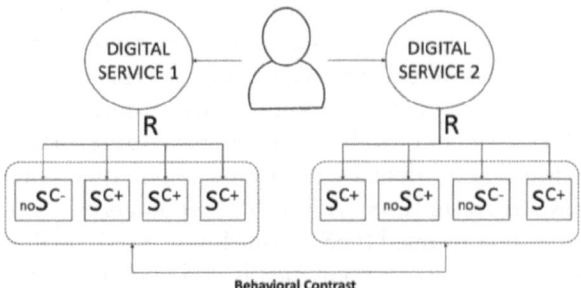

Figure 7.2 Behavioral contrast between two digital services
This figure exemplifies the concept of behavioral contrast between two digital services. In an already established digital behavior, changing the value of one of the reinforcers initiates competition with the available reinforcers from other digital services. Furthermore, the digital behavior designer must consider that not only the value associated with the reinforcers but also other variables such as the timing of administration, the reinforcement schedules through which the reinforcers are delivered, the drives, the strength of habit, among others, can affect the decision-making process regarding the use of the digital service.

individual will tend to increase the performance of the other behaviors (unchanged components). This is known as positive behavioral contrast, exemplified by the increased use of Telegram following the outage of WhatsApp. Conversely, if behavioral change occurs because the reinforcement of one behavior is altered in a parameter that positively modifies its value (changed component), the individual will tend to decrease the performance of the other behaviors (unchanged components). This is referred to as negative behavioral contrast, illustrated by the case of George increasing his Instagram usage due to a positive change in one of its reinforcers (changed component), thus reducing his use of other social networks (unchanged component). A well-known, real-life instance of negative behavioral contrast was the introduction of a new reinforcer in social media, the Stories feature. In 2013, Snapchat introduced Stories, a new functionality of videos that only lasted online for twenty-four hours and could be shared among friends. This feature achieved remarkable success with Generation Z and Millennials, and brands found this to be an ideal medium for sharing news and updates with their followers. Chris Cox, then the Chief Product Officer of Facebook, remarked at a conference that "the increase in the Stories format is on a path to surpass feeds as the primary way people share things with their friends sometime next year." The swift rise in the usage of Stories did not go unnoticed by other social networks, which, by 2017, had ended up "copying" and incorporating Snapchat's Stories into their services.

Indeed, they were not copying the Stories themselves but rather the reinforcing power of them. According to a 2018 report by the marketing agency "Block Party," Stories were shared at a rate up to 15 times higher than regular posts, and in 2018, 970 million accounts posted a Story daily on Instagram, WhatsApp, Snapchat, and Facebook (TechCrunch 2018). This rapid growth is due to the fact that Stories have great value as reinforcers because they are quick and easy to create (noS^{C-}), fun (S^{C+}), spontaneous (no preparation; noS^{C-}), effortlessly consumed (noS^{C-}), quickly accessed (noS^{C-}), provide real and daily information about your friends and favorite personalities (S^{C+}), offer direct access to your followers (S^{C+}), are customizable (S^{C+}) and disappear after twenty-four hours (noS^{C-}) (reinforcement matrix); they also present some punitive aspects such as not being storable long-term, although this has been mitigated by the implementation of Highlights on Instagram. Nevertheless, the fact that Stories disappear after twenty-four hours also plays a significant role, as it intensifies the behavior of using the app due to a mechanism known as "limited hold," associated with reinforcement schedules (Black *et al.* 1972). That is, "limited hold" indicates that the reinforcer will only be available for twenty-four hours, in the case of Stories. This "limited hold" increases the reinforcing value of the Stories, thereby intensifying the social media users' response to obtain them. Due to these characteristics, Stories caused a behavioral competition between the behavior of uploading content through Stories and that of uploading regular posts to the feeds of social media. The potent reinforcing value of the Stories was the true reason why they were swiftly copied by all social media platforms.

7.5   Role of Reinforcement Schedules in Behavioral Contrast

A substantial body of experimental work has uncovered two intriguing characteristics of digital service utilization when temporal constraints are mixed with behavioral competition: (1) behaviors from different digital services compete for the finite time of users, and (2) there are regulatory behavioral processes that tend to maintain user activities at a fixed proportion (Staddon 2003; Waltz & Follette 2009). The key to predicting the time a user will spend on one digital service over another is primarily determined by the administration of reinforcers, with intermittent reinforcement being more potent than continuous reinforcement (Ferster & Skinner 1957). This concept, linking the administration of reinforcers with their response rate or digital service usage rate, is known as "matching" (Davison & McCarthy 2016). Behavioral matching describes a mathematical relationship between the time spent engaging with a digital object and the reinforcement ratio obtained from this behavior. That is, the time spent on a digital service is related to the number and value of reinforcers the user receives from it in relation to those they would receive from all other latent behaviors they could be engaging in. Nonetheless, the user must exhibit some degree of behavioral variability for behavioral contrast and shaping to occur (Page & Neuringer 1985). This means that the user must know how to use the other services, suggesting that education in the use of digital services should be one of the initial strategies to compete with other behaviors. For behavioral contrast to occur, and thus, an increase in time and frequency of use of the new digital service, certain guidelines should be followed. For instance, in the design of reinforcers to shift a user from one digital service to another, the "resistance to change" due to an already established habit is critical (Poldrack 2021). The perceived value of the reinforcer by the potential new user must be significantly greater than that received from the habitual behavior for the change to occur (Kahneman & Tversky 2018), although there are different strategies to weaken existing habits (Bouton 2021). Another example is that the marketing strategy should not be directed at suppressing the current behavior of the user, as this type of strategy can lead to a paradoxical effect known as "rebound effect" and increase its use (Wegner *et al.* 1987). In behavioral change for substituting one habit for another, the same associative processes involved in habit formation are utilized (Berman & Dudai 2001; Bouton 2000), hence the solution does not lie in suppressing the old habit, but in facilitating or generating a new behavior that competes with the usual one (Horváth *et al.* 2022).

Some key strategies for facilitating this behavioral transition toward the desired behavior involve associating the new behavior with contexts relevant to the user, providing retrieval cues, and subsequently, generalizing the new behavior to new contexts (Bouton 2000). For example, consider a new start-up known as Health-gizer, which aims for its users to adopt healthy living habits by reviewing their daily actions, identifying unhealthy behaviors, and setting small goals to change these behaviors. Marie, the company's digital behavior designer, has prepared a robust bank of available reinforcers, with her molar reinforcer being improved health (subjectively assessed by a daily survey), implying better quality and longer life. Marie is aware that she competes with other health apps but decides that the best context for using her app is before going to sleep. She also considered that the morning could be a better context, but the competition with other reinforcers and punishments was significant, such as being late for work. Once the context is decided, Marie must design a set of discriminative stimuli that will act as retrieval cues, meaning, reminders for the user that the app exists and to use it. For this purpose, she designed certain notifications (discriminative stimuli) that appear before bedtime showing different behaviors one can do with the app and the reinforcer obtained, such as a healthy, easy, and economical breakfast, walking 1,000 steps, and so forth. This is possibly the most crucial step for any technological start-up: prompting the user to remember that the service exists at the right time. After achieving that users review the app before going to sleep, Marie has devised a strategy to generalize the behavior of consulting the app in various contexts by sending notifications in the morning or at different times of the day reminding them of the commitments made at night to improve their health. However, these users already have established behaviors before going to sleep at night, such as checking social media. At Health-gizer, the marketing department suggests that the best option is to suppress the behavior of checking social media at night through an aggressive campaign about the harm of social media to mental health. In contrast, Marie leans toward creating other types of reinforcers to compete with the behavior of viewing social media at night. These reinforcers could include, for instance, creating a social community that allows Health-gizer users to share their health care achievements and/or allowing them to share their achievements on other social networks to obtain social reinforcers such as congratulations from other users. This latter strategy also serves to reach out to new users. According to theory, the strategy of offering alternative reinforcers is better than suppressing behaviors (Horváth *et al.* 2022). Lastly, it is crucial to ensure that, before launching any plan to induce behavioral contrast in potential users and to compete behaviorally, the administration of new reinforcers is stable. That is, ensuring that the technological service is technically capable of receiving users so that the supply of reinforcers is not interrupted. Discontinuing the administration of reinforcers during behavioral competition can be catastrophic because users might revert to their previous habits (John *et al.* 2011; Silverman *et al.* 2012). The core of innovation for the digital behavior designer means developing and implementing a new reinforcer or modifying an existing one from a digital technological tool, in such a way that it sufficiently changes in value for the user, thus provoking a behavioral contrast with the services of the competition.

CHAPTER 8

Methodology for the Design of Digital Behaviors

8.1 Scope of the Methodology

Understanding the psychological foundations of user behavior in digital environments is imperative to design digital products and services. The aim of developing these technologies is for the digital behavior designer to identify and structure the basic functional elements that can guide users in performing different digital actions. In essence, this involves constructing a digital operant box. It is also vital to remember that the design is not aimed at ensuring 100 percent of users exhibit a specific behavior, but rather at increasing the percentage of a desired behavior in relation to a baseline, which should be established before designing, developing, and implementing changes in the digital service. Digital behavior designers essentially work to anticipate behavioral trends. Once these elements that will control the user's digital behavior are identified, the user experience and the interfaces of the digital service are designed around this knowledge. Specifically, the digital behavior designer develops behavioral blueprints of the user interacting with the digital service, focusing on personal, environmental, historical, and digital characteristics that modulate the user's behavior in the digital world. Personal characteristics refer to the user as a system embedded in a physical environment with needs to be satisfied; historical characteristics refer to users as systems that accumulate experiences generating expectations of using the digital service; and digital characteristics refer to the digital stimuli of the technological service that will signal which interactions must be performed to achieve a specific reinforcer. In particular, users can be operationalized as open learning systems of organic and computational basis capable of anticipating future events through processing contextual information associated with past experiences, thereby generating future representational expectations that control their behavior in a way that subjectively maximizes their benefit by solving physiological and psychological needs through digital tools. The discrepancy between expectation and reality is what would be considered learning and would modify the neuronal representations in which past experiences are encoded. Therefore, knowledge from the field of cognition and behavior will be used to anticipate how and why the user will interact with the digital service.

Next, a methodology will be proposed by which to identify, select, and prioritize the appropriate reinforcers to satisfy the drives of potential users of the technological service. This process is known as "digital behavioral design" and consists of a sequence of two phases:

1. Digital behavior analysis: detection of the functional elements that control molar behavior for the use of a digital service.
2. Drives and operants design: establishment of functional relationships between drives and satiating reinforcers.

These phases are consecutive, intrinsically related, and divided into different stages each (see Figure 8.1).

The digital behavior analysis phase is divided into an initial stage for the detection of the main drive of use and the associated molar behavior (O^D),

DIGITAL BEHAVIOR DESIGN

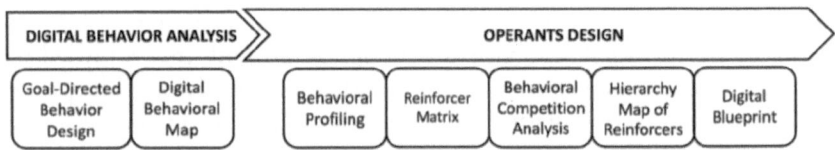

Figure 8.1 Phases for the development of "digital behavioral design"

Digital behavioral design is divided into two phases, which are further subdivided into several stages. The first phase consists of digital behavior analysis, which involves the design of the principal digital behavior that will support the digital service (goal-directed behavior design), and an initial scheme of its functionality (digital behavioral map). The second phase is a continuation of what was achieved in the first and is called "operants design." This phase is divided into the following stages: (1) behavioral profiling, which involves the identification of competing behaviors; (2) reinforcer matrix, which involves the identification of reinforcers for the competing behaviors; (3) behavioral competition analysis or the assignment of value to the different reinforcers of the various behaviors; (4) hierarchy map of reinforcers, which involves the ordering of the reinforcers that can be implemented in the digital service according to the total assigned value; (5) digital blueprint, which involves proposing an initial scheme of the digital service. Additionally, there is a final phase known as postponed digital features analysis, in which all features not considered in the initial digital blueprint will be stored.

and a second stage for the design of the functional elements of the digital service's molar behavior. The drives and operants design phase consists involves conducting a behavioral profiling of a standard user as third stage, the fourth stage of developing a matrix of reinforcers, the fifth stage of prioritizing the order of implementation of the different reinforcers, and the sixth and final stage, for assigning each reinforcer an appropriate reinforcement schedule. The following sections will explain each of these stages in more detail.

8.2 Digital Behavior Analysis (DBA)

DBA is the phase in which the molar behavior of the digital service is identified and designed. To achieve this, one must initially have a concept in mind for the digital service they wish to develop. Various methodologies exist for idea generation, but the principal one is that it should solve a problem or improve an existing process of a previously solved problem. From a psychological perspective, detecting where problems lie is accomplished by identifying the sources of human emotions and what triggers them. For instance, an event that induces anger or sadness suggests a process not aligning with people's expectations. Hence, it is a process that needs resolution through some mechanism (product) or service. Similarly, feelings of joy or euphoria might signal an event meeting a person's expectations, indicating a process that could inspire a business idea. Here, the aim is to select a viable business idea beneficial to the user. Once a digital business idea is conceived, it is vital to determine its correctness. Typically, the first step involves conducting a brief benchmark and business model analysis, simply to ensure the idea's validity. It is often nonintuitive to develop the business model prior to idea development, but frequently, after designing a digital service, when seeking investors or partners, the business model development may reveal unprofitability or the need for service modifications to achieve profitability. Therefore, it is advisable to approach this early on before delving deeply into service design, as this may uncover aspects crucial to the service design itself.

8.2.1 Goal-Directed Behavior Design

This stage of the DBA involves designing the potential contingencies occurring within a behavioral segment, defined as: $[\mathbf{O^D}: \mathbf{CTXT\text{-}S^D\text{-}P^B} \rightarrow \mathbf{R} \rightarrow \mathbf{S^C}]$ (see Table 8.1).

Table 8.1 *Goal-direct behavior design*

O^D	PRIMARY MOLAR BEHAVIOR				
	Ctxt	S^D	P^B	R	S^C
	SECONDARY MOLAR BEHAVIORS				
	Ctxt	S^D	P^B	R	S^C

A table for identifying the functional elements of primary and secondary molar behaviors that may occur when a user is in a determined drive state.

This table is an adaptation of the classic table used in the functional analysis of behavior by clinical psychologists. The main difference is that functional behavior analysis typically examines behaviors already performed by the person. In contrast, DBA anticipates user behavior. The first step in completing this table is understanding the functional relationships between the various functional elements. In DBA, the contingency relationship presented by $[\mathbf{O^D: CTXT\text{-}S^D\text{-}P^B} \rightarrow \mathbf{R} \rightarrow \mathbf{S^C}]$, can be divided into three functional blocks that will guide the design of users' digital behavior (see Figure 8.2).

The first functional block ($[O^D]$) refers to the user's drive that needs to be satisfied. This block will condition the next functional block, as the sensory stimuli present in the environment gain meaning in relation to the drive the user is experiencing. The second functional block ($[CTXT\text{-}S^D\text{-}P^B]$) is defined by a specific context where there are discriminative stimuli indicating the possibility of obtaining a reinforcing stimulus that could satisfy the drive. The user triggers different P^B upon perceiving an S^D, generating expectations about the reinforcer's value (goal-value representations). These expectations will condition the third functional block ($[R \rightarrow S^C]$), as initiating the action leading to the reinforcer will partly depend (but not solely) on the likelihood of successfully achieving the reinforcer. Lastly, the third functional block shows the functional relationship between digital behavior and the acquisition of the reinforcer that satisfies the drive. A key

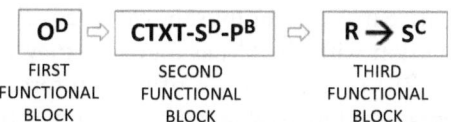

FIRST FUNCTIONAL BLOCK   SECOND FUNCTIONAL BLOCK   THIRD FUNCTIONAL BLOCK

Figure 8.2 Functional blocks of digital behavior analysis

The first functional block is the organism's drive (O^D). This drive is an internal vector generated through internal homeostatic changes indicating a physiological or psychological need that must be resolved. This drive conditions the relevant contexts for us, forming the second functional block of Digital Behavior Analysis. In this block, there is an almost reflexive relationship between how the context determines the discriminative stimuli (S^D) of value, which, when perceived, trigger prior beliefs (P^B), expectations, or anticipation of the possible reinforcers that could satisfy the drive. These PB can be defined as numerous mental representations of different $R{\rightarrow}S^C$. In this phase, the individual makes a decision about an R to carry out, which will lead to an S^C that satisfies the O^D. This final R-S^C is what is considered the third block.

point in developing a digital service is to match the reinforcer's expectations with the actual characteristics of the reinforcer, as this match will produce an emotion associated with the service and the likelihood of repeating the behavior.

Table 8.1 contains three major divisions: O^D, primary molar behavior, and secondary molar behaviors. In the O^D division, a single drive is positioned, which is considered to be able to lead the user to utilize the digital service being designed. Next, the second division is "the primary molar behavior," which refers to the user's main usage behavior. Lastly, the third division refers to "secondary behaviors to the primary molar behavior," meaning those actions that could complement the primary behavior and help satisfy the drive, taking into account certain particularities. For example, in a music service app, the primary molar behavior might be listening to music, and the secondary molar behaviors could include searching for favorite singers, saving songs, and so on.[1] Both primary and secondary molar behaviors are further divided into different columns, representing various elements of the DBA. It is the filling of these columns and the establishment of their contingency relations that is considered to be conducting a DBA. Specifically, these elements represent the second and third functional blocks described earlier. In the "Ctxt" column, the

---

[1] These secondary molar behaviors in one drive could turn out to be primary molar behaviors in other drives. For instance, in the drive for organization, searching for and saving songs in Spotify playlists might constitute the primary molar behaviors.

context where the drive may appear must be captured. The context must be analyzed in conjunction with the S^D, as together, they will determine the user's prior beliefs (PB) or expectations of the reinforcer. Specifically, P^B are mental representations (goal-value representations) about which behaviors to perform to obtain a reinforcer of a certain value. These P^Bs will be what the user cognitively processes for decision-making on which behavior to ultimately perform to satisfy the drive, whether to use the digital service or opt for another behavior that involves using another service. That is, the expectations that indicate the behavior and the value of the reinforcer will determine the suitability of initiating a behavior and not others to resolve the drive. Therefore, if the digital service's reinforcer does not meet the expected value or the behavior to be performed does not correspond to the value of the expected reinforcer, potential customers may choose a behavior that leads them to use another service with reinforcers closer to the expected. Therefore, it could be concluded that the main objective of a digital behavior designer is to design reinforcers with value for the different contexts in which potential users can find themselves. For example, the value of the reinforcer expected when your partner asks you "what are we going to eat today?" is very different from the value of the reinforcer expected when the same question is asked by a colleague. In both contexts, you expect the same reinforcer, but with different value. While with your partner, your expectations might be to make something quick to eat or more intimate, with your colleagues it could be to have a space to meet with them for social interaction. In both contexts, you expect to obtain food, but with different characteristics that will impact its perceived value. In the specific case of workers wanting to go out to eat, a new functionality of a digital service could be generated, such as inviting a friend to eat at a restaurant, thereby creating a loyalty system based on the number of times they go to eat and the average ticket (fixed ratio concurrent reinforcement program). The more money spent and the more frequent visits to the restaurant, the more reinforcers such as discounts can be obtained. The point is that it is through establishing and understanding the functional relationships of $[\mathbf{O^D}: \mathbf{CTXT}\text{-}\mathbf{S^D}\text{-}\mathbf{P^B} \rightarrow \mathbf{R} \rightarrow \mathbf{S^C}]$ that this new functionality has been designed. Next, a brief description of the functionality of elements from Table 8.1 will be provided. For further in-depth analysis, refer to previous chapters.

1. Drive (O^D): Anticipate the internal state of the individual that needs to be satisfied through the attainment of a reinforcing stimulus. Only one is analyzed per table. The recommendation when starting the

design of a new digital service is one drive for each dimension: self-preservation and psychological.

2. Context (Ctxt): Anticipate the context in which the user becomes aware of the drive, as it can determine the user's expectations about the value characteristics that the reinforcer must meet to satisfy the drive. It should be analyzed in conjunction with the S^D.

3. Discriminative Stimulus (S^D): Anticipate the specific stimuli that would signal to the user the possibility of satisfying a drive through the acquisition of a reinforcer.

4. Prior Belief (P^B): These are goal-value representations, which anticipate different properties of the reinforcer that will determine its value (R-S^C), which will be a crucial step in decision-making to carry out the behavior.

5. Operant Response (R): Anticipate the digital behavior that the user will have to perform to obtain the reinforcers.

6. Contingent Stimulus (S^C): Determine the reinforcers that could be obtained by performing the digital behavior, conditioned by the context, the discriminative stimuli, and the prior beliefs.

Next, each of these elements will be briefly explained within their functional blocks of the DBA. A theoretical justification for the selection of these elements can be found in the preceding chapters of this book.

8.2.1.1 First Functional Block: Selection Drive [O^D]

Regarding the design of the service itself, from the DBA perspective, the initial questions a designer should ask pertain to understanding the drive of the user that is being satisfied. For instance, Uber Eats addresses the self-preservation drive of hunger, while Twitter satisfies psychological drives like curiosity, social contact, acceptance, and/or status. A single service may address multiple drives for different individuals or numerous drives for the same person. Hence, it is crucial to also identify the primary drive that our digital service is addressing.

The detection of a service's primary drive is accomplished by identifying the various drives that the digital service caters to. The most effective way to identify this is through a table that reviews each drive to determine if it is being addressed (see Table 8.2).

In this table, the drives of self-preservation are deemed more significant than psychological drives, as these can overshadow the psychological drives (Pozo *et al.* 2023). However, there are propositions that intellectual security (feeling of self-esteem) is the main motivator of all human behaviors (Jaffe 2010), just as the possibility of obtaining information can suppress some self-preservation drives (Rodriguez Cabrero *et al.* 2019). Nevertheless, the drive of self-preservation should be prioritized.

Table 8.2 *Drive selection*

DRIVES	Type	Sub-type	Check	Why
Self-Preservation	Survival	Hunger		
		Thirst		
		Pain		
		Cooling		
		Sofocation		
		Urinary retention		
		Exhaustion		
		Sleep		
		Inactivity		
		Habitat		
		Hedonism		
	Sexual	Reproduction - Love		
		Offspring maintenance - Family		
Psychological	Social	Communicative behaviors		
		Social contact		
		Honor		
		Identity control		
		Status		
		Domination		
		Self-expansion		
	Meta-Cognitive	Curiosity		
		Revenge		
		Order		
		Idealism		
		Tranquility		
		Saving		
		Social acceptance		
		Physical exercise		
		Independence		
		Cognitive validation		
		Universalism		

A table for detecting the drives satisfied by the digital service.

8.2.1.2 Second Functional Block: Prior Beliefs [Ctxt – S^D – P^B]

It is essential to understand at this juncture that the **[Ctxt – S^D – P^B]** operate as a single functional unit. That is, S^D acquires meaning according to the context, and both elements (Ctxt-S^D) will trigger a series of Prior Beliefs, goal-value representations, mental representations, or expectations of the value of obtaining the reinforcer. In this sense, the Context is a variable that can condition the functional relationships of the other elements of goal-directed behavior. The behaviors of individuals are immersed in a specific environment, filled with sensory stimuli that trigger various mental representations or possible alternative futures in people. For the designer of digital behaviors, analyzing user behavior outside their immediate perceptual context is nonsensical. This context is determined by various factors such as prevailing emotional state, personality, and learning history. Therefore, defining the environment is a fundamental task to anticipate people's potential decisions. The more detailed this description, the better we can identify the reinforcer the individual needs to satisfy the drive in that context. Nevertheless, it is important to remember that the digital behavior designer does not design for a specific individual, but for behavioral tendencies, hence a recommendation to conceptualize a generic context rather than considering a specific emotional state, personality, and learning history, at least initially. On the other hand, the analysis of the context must be conducted in conjunction with S^D, as the contingencies following it, such as Prior Beliefs, are conditioned by the context. Asking "what are we going to eat today?" at work with colleagues is different from asking the same question at home with your partner. Likely, the meal at work would carry more social-origin drives such as social contact or status, compared to with your partner where it may be more related to self-preservation.

Stimulus Discriminative is the environmental stimulus which triggers the expectation of obtaining a reinforcer in a given context, acquiring meaning according to the context (Ctxt) where it is found. In the column of S^D, those sensory stimuli that trigger the expectations of obtaining a reinforcer to satisfy the current drive should be noted. For example, seeing a McDonald's sign when hungry can anticipate that entering, queuing, and paying for a burger may alleviate hunger. Similarly, photographs of food on the internet, checking mealtime on a cellphone, observing others with food, or being asked by your partner what we are going to eat today . . . all these are S^D, which in some way trigger P^Bs or expectations of obtaining food. The emotion associated with the drive will moderate the intensity and urgency of obtaining these foods. In this case, it is crucial to accurately define the drive and the context associated with the S^D, because the same S^D can trigger different P^B with different drives and in different contexts. For instance, the same S^D can act for two distinct drives. If you receive a dining invitation from your boss, this invitation can satisfy both the hunger drive and the status or dominance drive. While the hunger drive will be satisfied with the food, the status or dominance drive will depend on the person's expectations and the outcome of interactions during the meal with the boss. Furthermore, as previously explained, it is also important to describe the context in which the individual is in order to better refine the features of the technological service in terms of offering what the user truly needs.

In the P^B column, the digital behavior designer must be aware that at this point, the user has not yet performed the behavior, but is mentally and subjectively anticipating the value of obtaining this reinforcer. The term coined for the anticipation of the characteristics (and therefore, the value) of the reinforcer to be obtained is "prior belief," which is generated from the user's previous experiences with the reinforcer, or with other similar services, or even with references from other users. Therefore, this column should reflect those characteristics of the reinforcer that are vital to maintain its value when the user obtains it. In a way, P^Bs condition how the reinforcer should be designed and administered to fully satisfy the user's drive. Identifying these value-adding characteristics of the reinforcer focuses corporate efforts on providing the user with what they expect, thereby creating an emotional association with the brand, service, and products. Below, a series of utilitarian dimensions that a reinforcer can acquire to endow it with value will be shown (see Figure 8.3).

• Functional value: Refers to the perceived value obtained from the ability of an option to perform useful, practical, or physical functions. An option acquires functional value through the possession of outstanding functional, useful, or physical attributes. In the case of food, its functional value would be the caloric intake to eliminate hunger.
• Conditional value: Refers to the perceived utility obtained as a result of the specific situation or circumstances prior to making a decision. An

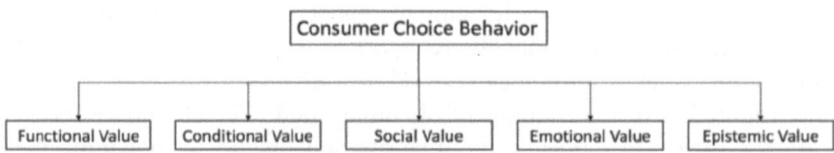

Figure 8.3 Functional dimensions for assigning value to reinforcers
Proposals on different hedonic characteristics that can add value to stimuli for users
(Sheth *et al.* 1991).

option acquires conditional value in the presence of preceding physical or social contingencies that increase its functional or social value. For example, going to eat at a restaurant takes on conditional value if you are accompanied.

- Social value: Corresponds to the perceived utility gained by associating an option with one or more specific social groups. An option gains social value through its connection with demographic, socioeconomic, and cultural-ethnic groups. In the case of food, its social value is when it is shared with your friends.
- Emotional value: This is the perceived utility obtained through the ability of an option to evoke feelings or emotional states. An option gains emotional value when it is associated with specific emotions or when it provokes or perpetuates such feelings. In the food example, the emotional value would be that food which reminds us of our childhood meals or those made by our grandmother.
- Epistemic Value: Refers to the perceived utility gained through the ability of an option to stimulate curiosity, provide novelty, and/or satisfy the desire for knowledge. In the case of food, it would be discovering new food that presents some characteristic that we evaluate as reinforcing. For example, discovering foods from other countries that one has never tasted.

When a user receives food at their home, driven by hunger, the P^B is functional, as the individual expects a reduction in hunger. How can a company ensure that the reinforcer reduces hunger? Such decisions are corporate-level, but can be central to product development. A hypothetical food delivery company might offer a line of high-calorie foods or large portions. Flavor is another crucial aspect, though it satisfies the hedonism drive. In the final design of the digital service, it should satisfy not just one drive but multiple drives simultaneously, increasing the value of the reinforcer. These expectations are linked to a S^D and context, triggering the P^B of the available reinforcer and behaviors needed to obtain it. For instance, when a potential user of a future digital service is asked by their partner at home what they will eat today, generally two prior beliefs may arise: "need to cook" or "must have enough food for two." The potential user, facing this situation, seeks a reinforcer that satisfies the drive (food), but with these conditions attached to the value of obtaining the food. Similarly, to the same question in a work context with colleagues, the conditions imposed by the P^B on the reinforcer determine that besides food, it is crucial that it be served in a place where they can comfortably gather outside the office, possibly with an option to reserve in advance to ensure availability or proximity to their workplace. As demonstrated in this example, the same drive, the same S^D, can generate different P^B due to the context, and these P^B will condition the value of the expected reinforcer.

8.2.1.3   Third Functional Block: Final Value-Decision Making [R → S^C]

Following the identification of the functional relationship of $Ctxt - S^D - P^B$, it is necessary to associate the molar operant response that the user must perform to obtain the anticipated reinforcer. This operant response should be the one preceding the acquisition of the reinforcer. For instance, if the food ordered through a digital service is the reinforcer, the operant response should be "pressing the confirm order button"; or if music is the reinforcer, the response should be "pressing the PLAY button of the song." This point is crucial, as it allows the digital behavior designer to apply an operant technique known as "backward chaining" to design the molecular behaviors that the individual must perform to reach the screen with the confirm order button. Reinforcers are placed in the last column of Table 8.1 (S^C). A technique that can be used to guide the identification of potential reinforcers is to follow the schema presented in Figure 8.4.

To identify reinforcers, it is advisable to explore those that can be obtained using the technological service in the short, medium, and long term. Within each category, a tracking of reinforcers that may appear with immediacy (a few seconds) or delay (minutes to days), with a positive or negative contingency relationship, and finally, whether there is direct or indirect mediation, that is, whether it is the technological device that administers the reinforcer or an external system to it, should be conducted. Once the different obtainable reinforcers have been identified, the main one should be determined. The two main rules that should guide the selection of the primary reinforcer are:

• Greater power to satiate the drive.
• More contiguous to the operant response.

The reinforcer with the greatest power to satiate the drive should be selected as the primary one, as this is the main reason why the user accesses the digital service. Moreover, this satiating reinforcer must be administered as soon as possible (contiguously) to the user to reduce the discounting effect, that is, the decay of the reinforcer's value due to the delay in its administration. For example, in a digital food delivery service, the main reinforcer is the food, although it takes a few minutes to administer; while in Spotify, it is the music, with its administration being instantaneous.

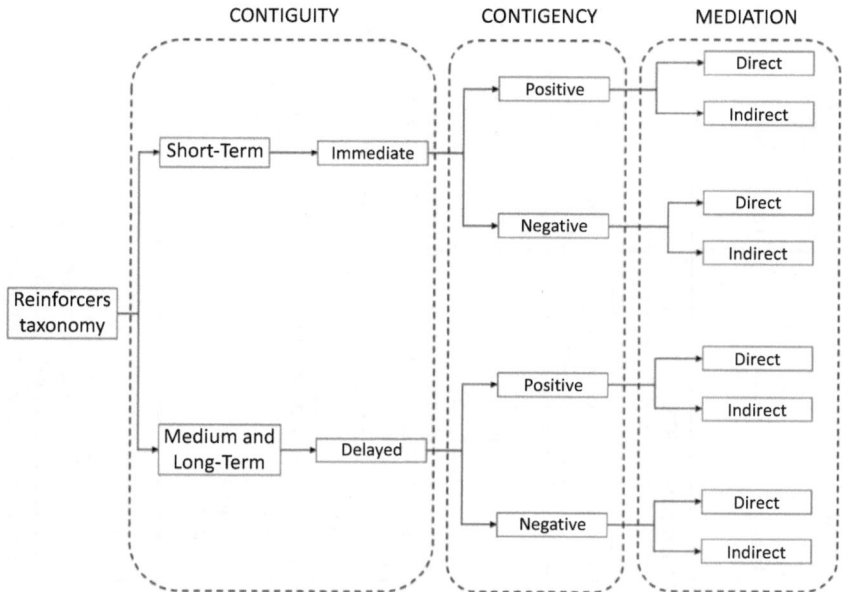

Figure 8.4 Technique for guiding the design of potential reinforcers (reinforcers taxonomy)

Reinforcers can be designed by following the different dimensions that indicate how they can be administered. There are 8 possible types of reinforcers: (1) Immediate Direct Positive Reinforcer (S^{C+id}); (2) Immediate Indirect Positive Reinforcer (S^{C+ii}); (3) Delayed Direct Positive Reinforcer (S^{C+dd}); (4) Delayed Indirect Positive Reinforcer (S^{C+di}); (5) Immediate Direct Negative Reinforcer (noS^{C-id}); (6) Immediate Indirect Negative Reinforcer (noS^{C-ii}); (7) Delayed Direct Negative Reinforcer (noS^{C-dd}); (8) Delayed Indirect Negative Reinforcer (noS^{C-di}).

8.2.2 Digital Behavioral Map

At this stage of the digital service design, all detected behaviors, both primary and secondary, should be documented. To illustrate how to create this map, Table 8.1 will be used for the design of a food delivery digital service. A drive of self-preservation and survival, hunger (see Figure 8.5), has been selected. Other drives such as hedonism, offspring maintenance, status, could be considered, but to simplify the work and its understanding, only the hunger drive will be analyzed.

Once the table is completed, a map of digital behavior, which will document all detected behaviors in the previous goal-directed behaviors design, must be created. The following observations are recommended for creating this map:

PRIMARY MOLAR BEHAVIORS

O^D	Ctxt	S^D	p^B	R	S^C
	Being alone at home without the desrise to cook.	- Clock indicating mealtime. - Empty fridge. - Fridge without desired food, visualizing snacks.	- Want to find the food I wish to eat. - Crave tasty food. - Want the food to be quickly to my home.	- Order food through an app.	Variety of food from different restaurants (S^{C+}). Food at home quickly (noS^{C-}).

SECONDARY MOLAR BEHAVIORS

	Ctxt	S^D	p^B	R	S^C
HUNGER	Being with your partner at home and it's time to eat.	- Your partner asks "what are we going to eat?"	- It must be to the liking of both. - Economically priced. - Ideally, we can split the payment so it doesn't all fall to me.	- Select food from various restaurants. - Select a shared menu. - Split the payment. - Discounts.	- Food from different restraunts (S^{C+}). - Discounts for ordering food together (S^{C+}). - Less money spent by splitting the bill (noS^{C-}).
	Having to eat alone at work.	- Break time and nothing prepared to eat.	- Must order in advance for the food to arrive. - I feel like eating out there. - I crave homemade food.	- Schedual food. - Reserve a table. - Choose homemade food filter.	- Preferred food instantly (S^{C+}). - Being able to reserve in a restaurant bar to eat (S^{C+}). - Option of homemade food (S^{C+}).
	Having to eat at work with colleagues.	- Going to eat with your colleagues.	- Let's see if we have a place where we can be comfortable. - Hopefully, wherever we go has what I like and it's cheap.	- Register a favorite restaurant.	- Reserve a spot near work (noS^{C-}), well-known (S^{C+}), and pleasant (S^{C+}).

Figure 8.5 Table of a digital behavioral map

This example has been made only with the hunger drive and considering that the main behavior is ordering food through a mobile app. A main context has been identified (being alone at home without the desire to cook), and four possible alternative contexts where hunger could also appear. Subsequently, this context has been associated with possible discriminative stimuli that indicate the available reinforcers. With the main behavior already defined, the prior beliefs identify the attributes of the ideal reinforcer to produce a match between expectation and reality. These prior beliefs depend on the contexts and discriminative stimuli. The operant responses are adjusted to what could be done to satisfy the prior beliefs. Finally, the contingent stimuli are the reinforcers that users would obtain if they perform these responses.

- Establishing the beginning and end of digital behavior. The initial behavior always starts by entering the digital service, and the final behavior usually coincides with the operant response of the main behavior, which in this case is pressing the "confirm order" button in the food app. Secondary behaviors must be arranged between the initial and final behavior of the technological service.
- All secondary behaviors are extracted from the table and attempts are made to group behaviors that belong to the same category or that are subordinate to one another. For example, "registering a favorite restaurant" might be subordinate to "viewing a restaurant," as when a preferred restaurant is not found, the user can register it on the platform.
- Once the groups are formed, try to put a label that indicates a meta-property of the grouped behaviors. In the example shown in Figure 8.6, these would be the labels of "restaurant selection," "way to administer reinforcement."
- Finally, arrange the groups of behaviors according to how they should appear in the app. A simple technique for organizing the different steps of the user until reaching the main molar behavior is the "funnel technique," which proposes going from the most general aspects to the more specific aspects.

The transition from goal-directed behaviors to a Digital Behavioral Map (DBM) can be seen below (see Figure 8.6).

In this digital behavior map, various phases can be visualized, which are determined by a meta-property (e.g., restaurant selection, way to administer reinforcement, social aspects . . .). It can be highlighted in bold which behaviors should be programmed first to develop a proof of concept. In the previous example, entering the app, viewing a list of restaurants,

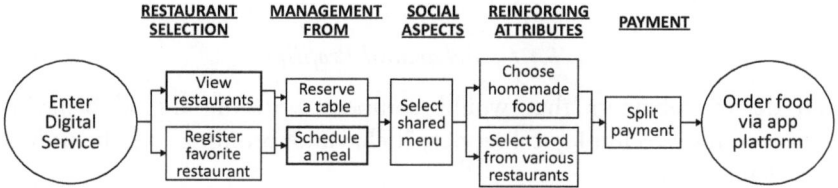

Figure 8.6 Example of a digital behavioral map
Example of the first Digital Behavioral Map obtained from the Goal-directed Behavior Design. This is a first proof of concept of the digital service and the different behaviors that can be performed. This is not its final version, as it will be refined in phase 2.

scheduling the meal, and ordering it is sufficient. Other features such as registering in the service or other configurations can be implemented in the proof of concept or wait for the minimal viable product. This proof of concept will be the main structure upon which subsequent features are implemented. It is noteworthy to mention that each of these behaviors appearing in the user's digital behavioral map in the technological service are of a molar type. Molecular behaviors are those that have to be performed to complete each of these molar behaviors.

Finally, one of the activities that can be performed in this phase to promote the creation and/or digital innovation is to change the character-istic order of digital behaviors. For instance, in Figure 8.6, all food services start by choosing the restaurant. What would happen if in a food service instead of choosing the restaurant, you directly chose what you want to eat? That is, placing the food choice first, and the app directly offering you the options you have. Furthermore, in this scenario, if several people ordered food from different restaurants, artificial intelligence algorithms could be applied to choose the restaurants based on proximity and workload to offer the best option for all dishes from different restaurants to arrive within the same time range. Therefore, the design of digital behavior prior to devel-opment is an activity that can determine new features of your digital service that will differentiate it from other services.

8.3 Drives and Operants Design (DOD)

The objective of this phase is to design the reinforcers that could be contingently implemented for the various molar behaviors shown in the digital behavioral map. To design optimal reinforcers, it is necessary to detect those behaviors that compete with the digital behavior of the service being designed. The technique to be used to detect these behavioral competitions is behavioral profiling.

8.3.1 Behavioral Profiling

To detect behaviors that would compete with the digital behaviors designed for a technological service, "behavioral profiling" will be utilized (see Table 8.3).

Behavioral profiling helps us detect similar behavioral patterns among potential users. When starting the design of a new technological service, behavioral profiling must be carried out for the most standardized behaviors. Minority or less frequent behaviors, unless specifically designing a service to

Table 8.3 *Behavioral profiling table*

Awakenings time		Fall asleep time
Primary Digital Behavior		
Context		
Behaviors		

Table for identifying competitive behaviors of the main digital behavior.

address them, will be left for other stages of development. Continuing with the previous example, to perform behavioral profiling, one must:

1. Select the potential target user.
2. Establish the time at which the average user wakes up and goes to bed.
3. Define the temporal space being analyzed: one hour, twenty-four hours, one week, etc.
4. Identify at what time of the day/week/month it is plausible that the target behavior will occur.
5. Identify possible contexts that may accompany the target behavior.
6. Identify potential competing behaviors.

The target behavior corresponds to the behavior that is intended to be established, while the competing behaviors are other behaviors that the potential user might perform to satisfy the drive. Continuing with the example of the digital food delivery service, the following figure is obtained (see Figure 8.7).

In Figure 8.7, the target user is selected as an average working single person. Rather than focusing on the personal characteristics of the target user, the digital behavior designer focuses on the potential behaviors this person might engage in, which are determined by environmental and personal factors. The analysis time for this behavioral profiling was twenty-four hours, as this is a sufficient temporal space to display different situations for satisfying the drive being addressed. In the context, the main behavior that the average target user might be performing is simply placed, while in the row of competitive behaviors are the alternative behaviors to using the

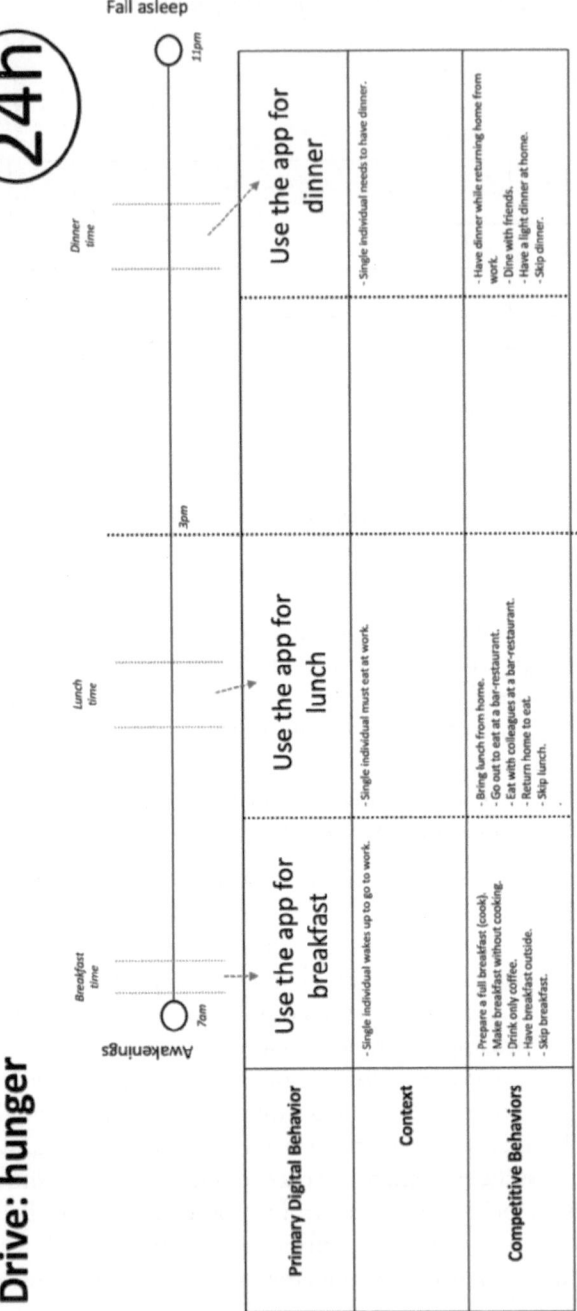

Figure 8.7 Example of behavioral profiling

An example of Behavioral Profiling for a food delivery service. Behavioral profiling is used to identify the most appropriate time period for the user (which can be assessed in days, weeks, months, or even years) to use the service. Once the temporal period is determined, it is crucial to define the context (usually the one proposed in the main behavior during the Digital Behavioral Map stage) and then identify which other behaviors might compete with the digital behavior being designed.

digital service that the target user might perform. To continue with the design of the operants, the digital behavior designer must choose which time segment to address for their analysis. The maxim at this point is: "Before being the best at everything, become the best at something." In this example, lunch time has been selected as the niche that the digital service wants to occupy. To successfully become a predominant digital behavior in this niche, a competition behavioral analysis will be carried out.

8.3.2 Reinforcer Matrix

The reinforcer matrix involves analyzing the reinforcers and punishments of alternative behaviors in order to design reinforcers that neutralize their effect in a behavioral contrast. In this way, our digital behavior can compete with the rest of the behaviors that a user might perform to satisfy the drive. Moreover, the reinforcer matrix can also be used to break habits of the potential user. To carry out this reinforcer matrix, the following table will be used (see Table 8.4).

The following section will describe the elements of the table:

- S^D: A stimulus indicating that the reinforcer is available.
- R: Behavior to be performed to obtain the reinforcer.
- S^C (description): A consequent stimulus contingent and contiguous to R.
- Type (consequent): A reinforcer when an appetitive stimulus follows R with sufficient power to increase the likelihood of R being repeated under the same S^D. Conversely, it will be a punishment if an aversive

Table 8.4 *Reinforcer matrix*

[TEMPORAL PERIOD]						
S^D	R	S^C Description	Type Consequent	S^C Contingency	S^C Contiguity	S^C Mediation

Table to identify the different reinforcers of the competitive behaviors.

stimulus follows R with enough power to decrease the likelihood of R being repeated under the same S^D.

- S^C (contingency): The contingency relationship between R and S^C. It will be positive when the likelihood of S^C occurring after R is high (likely to appear); and negative when the likelihood of S^C occurring after R is low (unlikely to appear).

- S^C (contiguity): The time elapsed between R and S^C. An immediate contiguity relationship will be considered when S^C appears almost simultaneously with R. In contrast, a delayed contiguity relationship will be considered when S^C appears delayed relative to R.

- S^C (mediation): Direct mediation occurs when the S^C is managed by the technology directly; whereas indirect mediation occurs when the S^C is managed by a system different from the technology used to obtain it.

Subsequently, a reinforcer matrix of hunger drive satiation at lunchtime for a single person with a job will be shown (see Table 8.5).

8.3.3   Behavioral Competition Analysis

After completing the reinforcer matrix, a behavioral competition analysis of the technological service being designed is conducted. The purpose of this behavioral competition analysis is to identify additional reinforcers to the digital behavioral map that allow the digital service to compete behaviorally with other services. For the creation of this framework, different reinforcers must be grouped, weighted, and prioritized using Table 8.6 (see Table 8.6).

8.3.3.1   Grouping Reinforcers

To group reinforcers, those identified in the Goal-Directed Behavior Design (Table 8.1) and Reinforcer Matrix (Table 8.4) should be compiled into a new table (see Table 8.6), removing duplicates and merging if any reinforcer or punishment is similar (see Table 8.7).

To identify potential reinforcers and transform them into digital features, one must think from the most obvious and simple (e.g., if the reinforcer is saving money, then display the price) to the more complex (e.g., earning points each time you buy). Other aspects to consider include synthesizing the information in Table 8.7. For instance, the "Money Saving" reinforcer appears in several tables, so it is only mentioned once in Table 8.7, while "Disconnecting from Work" and "Avoiding Work Stimuli" are similar reinforcers that could be merged into one such as "Work Disconnection (fewer work stimuli)." In Table 8.7, most reinforcers follow a continuous

Table 8.5 *Examples of a reinforcer matrix*

LUNCH TIME

S^D	R	S^C Description	Type Consequence	S^C Contingency	S^C Contiguity	S^C Mediation
Eating Time at Work	Eating Home-Cooked Food	I save money	Reinforcer	Negative	Delayed	Indirect
		I eat what I like	Reinforcer	Positive	Inmediate	Direct
		I don't gain weight	Reinforcer	Negative	Delayed	Direct
		I avoid social interactions	Reinforcer	Negative	Immediate	Indirect
		Home-cooked food	Reinforcer	Positive	Immediate	Direct
		I socialize less	Punishment	Negative	Immediate	Indirect
		I have to prepare the food beforehand	Punishment	Positive	Delayed	Indirect
		Wash the dish afterwards	Punishment	Positive	Inmediate	Indirect

LUNCH TIME

S^D	R	S^C Description	Type Consequence	S^C Contingency	S^C Contiguity	S^C Mediation
Eating Time at Work	Eating Alone in a Restaurant	Obtaining food	Reinforcer	Positive	Immediate	Directa
		Variety of foods	Reinforcer	Positive	Immediate	Directa
		Time for oneself	Reinforcer	Positive	Immediate	Indirect
		Meeting new people	Reinforcer	Positive	Delayed	Indirect
		Opportunity to read or work	Reinforcer	Positive	Immediate	Indirect
		Less work-related stimuli	Reinforcer	Negative	Immediate	Indirect
		Little interaction	Punishment	Negative	Immediate	Indirect
		No savings	Punishment	Positive	Delayed	Indirect
		Delay in service	Punishment	Negative	Inmediate	Directa

LUNCH TIME

S^D	R	S^C Description	Type Consequence	S^C Contingency	S^C Contiguity	S^C Mediation
Eating Time at Work	Eating with Colleagues at a Restaurant	Social interactions	Reinforcer	Positive	Immediate	Indirect
		Splitting the bill	Reinforcer	Negative	Immediate	Directa
		New contacts	Reinforcer	Positive	Delayed	Indirect
		Sharing experiences	Reinforcer	Positive	Immediate	Indirect
		Disconnecting from work	Reinforcer	Negative	Immediate	Indirect
		Taking advantage of promotions	Reinforcer	Negative	Immediate	Indirect
		Limited variety of food	Punishment	Negative	Immediate	Directa
		No savings	Punishment	Positive	Delayed	Indirect
		Potential conflicts	Punishment	Positive	Immediate	Indirect

LUNCH TIME

S^D	R	S^C Description	Type Consequence	S^C Contingency	S^C Contiguity	S^C Mediation
Eating Time at Work	Returning Home to Eat	Engage in a complementary activity to eating (watch TV)	Reinforcer	Positive	Immediate	Indirect
		Take a nap or lie down	Reinforcer	Positive	Immediate	Indirect
		Save money	Reinforcer	Negative	Delayed	Indirect
		Gain less weight	Reinforcer	Negative	Delayed	Directa
		Home-cooked food	Reinforcer	Positive	Immediate	Directa
		Spend time with partner/family	Reinforcer	Positive	Immediate	Indirect
		Avoid social interactions	Reinforcer	Negative	Immediate	Indirect
		Less time for eating (time to go home and return to work)	Punishment	Negative	Immediate	Indirect
		Less social interaction	Punishment	Negative	Immediate	Indirect
		Prepare food	Punishment	Positive	Immediate	Directa

LUNCH TIME

S^D	R	S^C Description	Type Consequence	S^C Contingency	S^C Contiguity	S^C Mediation
Eating Time at Work	Not Eating	Saving money	Reinforcer	Negative	Delayed	Indirect
		Avoid social interactions	Reinforcer	Negative	Immediate	Indirect
		More time to finish work	Reinforcer	Positive	Immediate	Indirect
		Intermittent fasting (e.g., for health)	Reinforcer	Positive	Delayed	Direct[a]
		Cognitive fatigue	Punishment	Positive	Immediate	Indirect
		No social interaction	Punishment	Negative	Immediate	Indirect

In this example, different competitive behaviors of the main digital behavior of a food delivery service were identified. The digital service aims to occupy the lunchtime space at work, hence behaviors such as "bringing food from home," "eating at a restaurant," "returning home to eat," or "not eating" are seen as competing behaviors to ordering food delivery. Each reinforcer and punishment are divided according to their contingency, contiguity, and mediation.

Table 8.6 *Behavioral competition analysis*

				Scoring				Total Points User Value
Type	Description	Digital Feature	Digita Reinforcement Programs	User Value	Development Time	User Value	Development Time	

Behavior Competition Analysis

A table for assigning value to previously detected reinforcers for both the digital behavior and competing behaviors. It is also necessary to associate each reinforcer with how it will transform into a digital feature of the digital service.

Table 8.7 *Example of behavioral competition analysis*

				Scoring				
Type	Description	Digital Feature	Digital Reinforcement Programs	User Value	Development Time	Development Cost	Logistical Cost	Total Points
S^{C+di}	Money Saving	Gamification: A system that allows for the exchange of points for food.	Reinforcement Schedule with a points requirement and a limited duration					
		Icon: price	Continuous reinforcement schedule without a set duration					
S^{C+id}	Variety of Foods (what I like)	Variety of different restaurants	Continuous reinforcement schedule without a set duration					
		Specialized Recommendations	Continuous reinforcement schedule without a set duration					
		Register Favorite Restaurant	Continuous reinforcement schedule without a set duration					
noS^{C-dd}	I Don't Gain Weight	Filter: Healthy food	Continuous reinforcement schedule without a set duration					
		Display Health Icon	Continuous reinforcement schedule without a set duration					
		Show Calories	Continuous reinforcement schedule without a set duration					

Behavior Competition Analysis

Table 8.7 (cont.)

| | | | | Scoring | | | | |
Type	Description	Digital Feature	Digital Reinforcement Programs	User Value	Development Time	Development Cost	Logistical Cost	Total Points
		Track Daily Calories	Continuous reinforcement schedule without a set duration					
noS^{C-ii}	Avoid Social Interactions	Filter: Menus with an excuse: simulating a calling activity.	Continuous reinforcement schedule without a set duration					
S^{C-id}	Homemade Food	Filter: Homemade food	Continuous reinforcement schedule without a set duration					
		Individuals Can Upload Their Own Dishes to the Platform for Others to Order	Continuous reinforcement schedule without a set duration					
S^{C-dd}	Meet New People	Discover Profiles: Encounter individuals requesting the same dish as you.	Continuous reinforcement schedule without a set duration					
		Meet People Visiting the Same Restaurant at the Same Time: Share the bill.	Continuous reinforcement schedule without a set duration					
S^{C-ii}	Opportunity to Multitask (like reading, working, watching TV)	Filter: Non-messy food (ability to continue your activities pleasantly while eating)	Continuous reinforcement schedule without a set duration					
noS^{C-ii}	Disconnect From Work (less work-related stimuli)	Filter: Friends' food (menu to share with friends)	Continuous reinforcement schedule without a set duration					

Behavior Competition Analysis

			Fixed-ratio reinforcement schedule with a time-based non-attendance requirement Variable-ratio reinforcement schedule.				
$S^{C\text{-}id}$	Social Interactions	Gamification: A map of connections with the people you dine with the most. Thus, the app can make special offers to dine with certain individuals you frequently eat with.					
$noS^{C\text{-}dd}$	Splitting Bills	Bill Splitting Option: Introduce an option to split the bill at the end. Enter the person's details, and they receive a message to contribute their payment.	Continuous reinforcement schedule without a set duration				
$S^{C\text{-}di}$	New Contacts	Special Promotion: The app offers a special promotion if you add a new contact and dine with them using the split payment option.	Continuous reinforcement schedule without a set duration				
$S^{C\text{-}ii}$	Sharing Experiences	User Experience: Allow users to have personal profiles and share their dining experiences. Avoid tying these experiences directly to the restaurant as it may penalize the business.	Continuous reinforcement schedule without a set duration				
$noS^{C\text{-}ii}$	Taking Advantage of Promotions	Utilize Promotions: Offer various promotions based on the total amount spent. Offering various promotions based on the amount of money spent at a restaurant.	Fixed-ratio reinforcement schedule with a digital service expenditure requirement. Fixed-ratio reinforcement schedule with a specific restaurant expenditure requirement.				

Table 8.7 (cont.)

			Behavior Competition Analysis					
				Scoring				
Type	Description	Digital Feature	Digital Reinforcement Programs	User Value	Development Time	Development Cost	Logistical Cost	Total Points
S$^{C \rightarrow id}$	Nap After Lunch or Rest When Going Home to Eat from Work	Filters: Foods that induce sleepiness	Continuous reinforcement schedule without a set duration					
S$^{C \rightarrow ii}$	Spending Time with Partner/Family	Filters: Family menu options	Continuous reinforcement schedule without a set duration					
S$^{C \rightarrow dd}$	Intermittent Fasting (e.g., for health)	Available Diets: Have diets available that support intermittent fasting	Continuous reinforcement schedule without a set duration					
noS$^{C \rightarrow ii}$	I Socialize Less When Eating Alone	Meet People Who Visit the Same Restaurant at the Same Time: Share the bill	Continuous reinforcement schedule without a set duration					
S$^{C \rightarrow di}$	I Have to Prepare Food in Advance (to take away or eat at home)	Filter: Morning meals (they are nutritious meals prepared to be consumed the following day)	Continuous reinforcement schedule without a set duration					
S$^{C \rightarrow ii}$	Washing Dishes Afterwards	Request: Ask for your food to be delivered in disposable containers	Continuous reinforcement schedule without a set duration					

S^{C-di}	I Don't Save Money	Gamification: A points system that you can exchange for food	Reinforcement schedule with point accrual and limited duration requirements.			
noS^{C+dd}	Service Delay	Ability to Schedule Orders: Have the option to program your orders	Continuous reinforcement schedule without a set duration			
		Communicate with Delivery Service	Continuous reinforcement schedule without a set duration			
		Filter: nearby restaurants	Continuous reinforcement schedule without a set duration			
		Filter: find your favorite restaurant	Continuous reinforcement schedule without a set duration			
noS^{C-dd}	Limited Variety of Food	Variety of Restaurants	Continuous reinforcement schedule without a set duration			
noS^{C+ii}	Potential Conflicts with Colleagues	Ability to Reserve Space in a Restaurant for Another Person and Their Guest: and leave a prepaid ticket	Continuous reinforcement schedule without a set duration			
noS^{C-ii}	Less Time to Eat (time to go home and return to work)	Filter: Meals that can be consumed in less than 10 minutes	Continuous reinforcement schedule without a set duration			

In this example, reinforcers detected in previous stages of Digital Behavioral Design have been primarily grouped, leaving the assignment of the value of each reinforcer blank for the moment.

reinforcement schedule without duration, although there are several excep-
tions. For example, the point system that can be exchanged for food follows a
reinforcement schedule with a point requirement and limited duration. This
means that for point acquisition, the user must purchase food through the app.
As they buy, points will accumulate that can be exchanged for rewards such as
discounts. The exchange of these points will have a limited duration (limited
hold) to motivate the user to use them quickly to foster habit formation. In
designing this point system, other consequential reinforcers of its implementation should be considered, such as being able to send points to your contacts or
redeem points for meals for friends as gifts. Another interesting reinforcer to
analyze is offering a discount promotion for those who used to eat together (for
booking at restaurants and/or splitting payments) but stopped doing so. For
example, a multiple reinforcement program could be designed, applying a fixed
ratio schedule with a time requirement for users who used to use the digital
service together. That is, if a pair of users regularly used the service, and then
stopped using it for a specified period, the service could generate and send a
special promotion to motivate them to return to using the digital service. If they
take advantage of the promotion, the multiple reinforcement program would
be reactivated, whereas if they do not, it could be discontinued. It is recommended that such programs be designed in conjunction with the company's
data and financial department to determine the timing and value of the
promotion. In this way, it does not upset the financial projections.

Once Table 8.7 has been completed, the next step is the assignment of value
to each reinforcer. This task has a high qualitative component, as the digital
behavior designer must accurately understand the vision and mission of the
company wanting to develop the service, the implications in terms of time and
programming difficulty, logistical costs involved in the implementation of the
reinforcer at the company level, and the potential cost of its development. In
this scenario, it is recommended to assemble a team from the company with
sufficient knowledge to perform an adequate weighting of the value for the
prioritization of these reinforcers in terms of their necessity and suitability.

8.3.3.2   *Reinforcer Value Assignment*

The valuation of each reinforcer will range from one to three, with one
indicating the least effort and three the greatest effort. Exceptionally, in
the "user value" category, this order is reversed, with a score of one being
of great value to the user and three of lesser value. However, some
reference elements that should be considered in the valuation of reinforcers in each of their dimensions will be presented below
(see Table 8.8).

Table 8.8 *Reinforcer value assignment*

	1	2	3
User Value	The reinforcer is immediate and direct.	The reinforcer is immediate and indirect.	The reinforcer is delayed.
Development Time	<1 month	1–6 month	>6 months (to be complex in programming and/or requires in-depth design)
Development Cost	Involves only one person from the team to implement the reinforcer.	Involves the creation of an internal group for its development.	Involves hiring or outsourcing external to the company.
Logistic Cost	Only involves the software development department. It is the development of a feature that does not require coordination with other departments or bodies.	Involves coordination with other departments or the implementation of an artificial intelligence solution.	Involves coordination with other departments and the implementation of an artificial intelligence solution.

Reference table for assigning value to the identified digital feature reinforcers during the behavioral competition analysis.

Through Table 8.8, a value is assigned to each possible feature, classifying it according to the effort required for its implementation. The example (see Table 8.9) simulates this task.

8.3.4 Hierarchy Map of Reinforcers

To fill out this table, a team was formed to evaluate each category that rated the total score of the digital features. Once the value is assigned to each category, each digital feature is ordered as having simple implementation effort (4–6 points), moderate implementation effort (7–9 points), and difficult implementation effort (10–12 points) (see Table 8.10).

Table 8.9 *Example of how value is assigned in the behavioral competition analysis*

			Behavior Competition Analysis	Scoring				Total Points
Type	Description	Digital Feature	Digital Reinforcement Programs	User Value	Development Time	User Value	Development Time	
$S^{C\text{-}di}$	Money Saving	Gamification: A system that allows for the exchange of points for food.	Reinforcement Schedule with a points requirement and a limited duration	3	2	2	2	9
		Icon: price	Continuous reinforcement schedule without a set duration	3	1	1	1	6
$S^{C\text{-}id}$	Variety of Foods (what I like)	Variety of different restaurants	Continuous reinforcement schedule without a set duration	1	1	1	2	5
		Specialized Recommendations	Continuous reinforcement schedule without a set duration	1	2	1	2	6
		Register Favorite Restaurant	Continuous reinforcement schedule without a set duration	2	2	2	3	9
$noS^{C\text{-}dd}$	I Don't Gain Weight	Filter: Healthy food	Continuous reinforcement schedule without a set duration	3	1	1	1	6
		Display Health Icon	Continuous reinforcement schedule without a set duration	3	1	1	1	6
		Show Calories	Continuous reinforcement schedule without a set duration	3	1	1	1	6
		Track Daily Calories	Continuous reinforcement schedule without a set duration	3	1	1	2	7
$noS^{C\text{-}ii}$	Avoid Social Interactions	Filter: Menus with an excuse: simulating a calling activity.	Continuous reinforcement schedule without a set duration	2	3	3	3	11
$S^{C\text{-}id}$	Homemade Food	Filter: Homemade food	Continuous reinforcement schedule without a set duration	1	1	1	1	4

Code	Theme	Description	Reinforcement schedule					
		Individuals Can Upload Their Own Dishes to the Platform for Others to Order	Continuous reinforcement schedule without a set duration	1	3	2	3	9
S^{C-dd}	Meet New People	Discover Profiles: Encounter individuals requesting the same dish as you.	Continuous reinforcement schedule without a set duration	3	2	2	2	9
		Meet People Visiting the Same Restaurant at the Same Time: Share the bill.	Continuous reinforcement schedule without a set duration	3	2	2	2	9
S^{C-ii}	Opportunity to Multitask (like reading, working, watching TV)	Filter: Non-messy food (ability to continue your activities pleasantly while eating)	Continuous reinforcement schedule without a set duration	2	1	1	1	5
noS^{C-ii}	Disconnect From Work (less work-related stimuli)	Filter: Friends' food (menu to share with friends)	Continuous reinforcement schedule without a set duration	2	1	1	1	5
S^{C-id}	Social Interactions	Gamification: A map of connections with the people you dine with the most. Thus, the app can make special offers to dine with certain individuals you frequently eat with.	Fixed-ratio reinforcement schedule with a time-based non-attendance requirement	1	3	2	3	9
			Variable-ratio reinforcement schedule					
noS^{C-dd}	Splitting Bills	Bill Splitting Option: Introduce an option to split the bill at the end. Enter the person's details, and they receive a message to contribute their payment.	Continuous reinforcement schedule without a set duration	3	2	1	2	8
S^{C-di}	New Contacts	Special Promotion: The app offers a special promotion if you add a new contact and dine with them using the split payment option.	Continuous reinforcement schedule without a set duration	3	2	2	2	9
S^{C-ii}	Sharing Experiences	User Experience: Allow users to have personal profiles and share their dining experiences. Avoid tying these experiences directly to the restaurant as it may penalize the business.	Continuous reinforcement schedule without a set duration	2	2	2	1	7
noS^{C-ii}	Taking Advantage of Promotions	Utilize Promotions: Offer various promotions based on the total amount spent.	Fixed-ratio reinforcement schedule with a digital service expenditure requirement.	2	2	1	2	7
				2	2	1	2	7

Table 8.9 (*cont.*)

Behavior Competition Analysis

Type	Description	Digital Feature	Digital Reinforcement Programs	Scoring				Total Points
				User Value	Development Time	User Value	Development Time	
		Offering various promotions based on the amount of money spent at a restaurant.	Fixed-ratio reinforcement schedule with a specific restaurant expenditure requirement.					
S^{C-id}	Nap After Lunch or Rest When Going Home to Eat from Work	Filters: Foods that induce sleepiness	Continuous reinforcement schedule without a set duration	1	1	1	1	4
S^{C-ii}	Spending Time with Partner/Family	Filters: Family menu options	Continuous reinforcement schedule without a set duration	2	1	1	1	5
S^{C-dd}	Intermittent Fasting (e.g., for health)	Available Diets: Have diets available that support intermittent fasting	Continuous reinforcement schedule without a set duration	3	1	3	3	10
noS^{C-ii}	I Socialize Less When Eating Alone	Meet People Who Visit the Same Restaurant at the Same Time: Share the bill	Continuous reinforcement schedule without a set duration	2	3	2	2	9
S^{C-di}	I Have to Prepare Food in Advance (to take away or eat at home)	Filter: Morning meals (they are nutritious meals prepared to be consumed the following day)	Continuous reinforcement schedule without a set duration	3	1	1	1	6
S^{C-ii}	Washing Dishes Afterwards	Request: Ask for your food to be delivered in disposable containers	Continuous reinforcement schedule without a set duration	2	1	1	2	6
S^{C-di}	I Don't Save Money		Continuous reinforcement schedule without a set duration	3	2	2	2	9

Code	Cost item	Digital feature	Reinforcer			
		Gamification: A points system that you can exchange for food	Reinforcement schedule with point accrual and limited duration requirements	3	1	6
$noS^{C\text{-}dd}$	Service Delay	Ability to Schedule Orders: Have the option to program your orders	Continuous reinforcement schedule without a set duration	3	1	9
		Communicate with Delivery Service	Continuous reinforcement schedule without a set duration	3	2	7
		Filter: nearby restaurants.	Continuous reinforcement schedule without a set duration	3	1	5
		Filter: find your favorite restaurant.	Continuous reinforcement schedule without a set duration	3	1	7
$noS^{C\text{-}dd}$	Limited Variety of Food	Variety of Restaurants	Continuous reinforcement schedule without a set duration	3	1	7
$noS^{C\text{-}ii}$	Potential Conflicts with Colleagues	Ability to Reserve Space in a Restaurant for Another Person and Their Guest; and leave a prepaid ticket	Continuous reinforcement schedule without a set duration	2	3	10
$noS^{C\text{-}ii}$	Less Time to Eat (time to go home and return to work)	Filter: Meals that can be consumed in less than 10 minutes	Continuous reinforcement schedule without a set duration	2	1	5

This figure illustrates how value is assigned to each digital feature-reinforcer listed in Table 8.7. This valuation is done following the instructions of Table 8.8, subsequently summing them into a total. This result will later be used to order each digital feature according to the effort of implementation in the digital service.

Table 8.10 *Hierarchy map of reinforcers*

Range	Implementation Effort Level	Explanation
4–6	Mild Implementation Effort	Easy and quick to implement
7–9	Moderate Implementation Effort	Moderately difficult and quick to implement
10–12	Major Implementation Effort	High difficulty and slow to implement
Mild Implementation Effort		
Reinforcer Description	Digital Features	Scoring
Moderate Implementation Effort		
Reinforcer Description	Digital Features	Scoring
Major Implementation Effort		
Reinforcer Description	Digital Features	Scoring

Table for ordering according to the implementation effort of the reinforcers and their associated digital feature.

The following will demonstrate an example of how Table 8.10 can be filled out (see Table 8.11).

Upon completing this section, the digital features, which are reinforcers that can satisfy the drive in different contexts for which users turn to use the technological service, must be integrated into a digital behavior blueprint.

8.3.5 *Digital Behavior Blueprint*

Blueprint is a term widely used in many technical areas, originally developed by British astronomer and photographer Sir John Herschel in 1842 as a cyanotype process (Ware 1999). This cyanotype process involved placing a

Table 8.11 *Example of hierarchy map reinforcers*

	Mild Implementation Effort	
Description of the Reinforcer	Digital Features	Score
Homemade food	Filter: Homemade food	4
Nap	Filters: Foods that make you sleepy	4
Opportunity to do other things while eating (like reading, working, watching TV)	Filter: Non-messy food (can do your activities placidly while eating)	5
Disconnect from work (less work-related stimuli)	Filter: Food with friends (menu to share with friends)	5
Being with partner/family	Filters: Family menu	5
Less time to eat (time to go home and return to work)	Filter: Foods that can be eaten in less than 10 minutes	5
Service delay	Filter: Find your favorite restaurant	5
Variety of foods (like what I like)	Variety of different restaurants	5
Not fattening	Filter: Healthy food	6
Not fattening	Show healthy icon	6
Not fattening	Show calories	6
I have to prepare food beforehand (to take away or return home to eat)	Filter: Foods for tomorrow (nutritious foods prepared for consumption the next day)	6
Service delay	Ability to schedule orders	6

	Moderate Implementation Effort	
Description of the Reinforcer	Digital Features	Score
Not fattening	Daily calorie count tracking	7
Sharing experiences	Users have personal profiles to share their dining experiences. Not tied to a specific restaurant as it could be a limitation for the business	7
Taking advantage of promotions	Offer various promotions based on the total amount of money spent	7
Service delay	Filter: nearby restaurants	7

Table 8.11 (cont.)

Moderate Implementation Effort

Description of the Reinforcer	Digital Features	Score
Splitting bills	Option to split bills at the end. Add a person and they receive a message to pay their share.	7
Washing dishes immediately	Request disposable containers for food delivery	8
Saving money	Gamification: A points system that can be exchanged for food.	8
Homemade food	People can upload their own dishes to the platform so that other users can order them	9
Meeting new people	Browse profiles that order the same dishes as you	8
Meeting new people	Meet people who go to the same restaurant at the same time. Option to split the bill.	9
Social interactions	Gamification: a connections map with people you	9
New Contacts	Special app promotion if you add a new contact and dine with them using the split payment option.	9
Service Delay	Communicate with the delivery service.	9
Variety of Foods (as per preferences)	Register your favorite restaurant.	9
Not fattening	Daily calorie count tracking	9

Major Implementation Effort

Description of the Reinforcer	Digital Features	Score
Potential Conflicts with Colleagues	The Ability to Reserve a Space in a Restaurant for Another Individual and Their Guest, Leaving a Prepaid Ticket (Seeking Forgiveness)	10
Intermittent Fasting (e.g., for better health)	Have diets available	10
Avoiding Social Interactions	Filter: Menus with an excuse: Simulated activity call	11

In this example, the reinforcers - digital features have already been organized according to the implementation effort required. The ranges have been: simple implementation effort (4–6 points), moderate implementation effort (7–9 points), and difficult implementation effort (10–12 points).

drawing on a semi-transparent sheet forcefully over a sheet of paper (often an architectural or electrical plan). This paper was coated with a photosensitive chemical mixture of potassium ferricyanide and ammonium ferric citrate, allowing the drawing from the semitransparent paper to be copied onto the paper with chemicals. The result was a copy of the original drawing with a blue background and the image traced in white lines. This way, in the past, numerous copies of plans could be made quickly. However, today, the use of the term "blueprint" has evolved to a more figurative meaning such as "detailed plan." In digital behavior design, digital behavior blueprint is used as the final plan with which UX and UI can start designing the digital service or product. The design of the digital behavior blueprint will be based on integrating the digital behavior map with the reinforcer hierarchy map in the following table (see Table 8.12).

In the leftmost column of Table 8.12, at the top, are the "Interfaces." These interfaces are initially the phases of the digital behavior map and correspond to the main screens through which users will communicate with the digital service to satisfy their drive. In contrast, digital behavior refers to the actions that a user can perform on these interfaces, some of which may enable other screens or pop-ups. On the other hand, digital features are the reinforcers that can be obtained on each interface. These digital features will be placed according to the effort of implementation. Finally, the postponed digital features refer to those digital features that have been decided to postpone because their implementation might exceed the capacity or suit-ability for the project's vital moment. Usually, the digital features that saturate in development time due to needing an elaborate design are considered as postponed digital features. The recommended instructions for filling out the digital behavior blueprint are as follows:

1. Place the interfaces found in the digital behavior map.
2. Place the digital behaviors that can be performed within the interfaces.
3. Place the digital features according to the required implementation effort (see Table 8.11). This point is important, as some reinforcers may not fit into any phase of the digital behavioral map, necessitating the inclusion of a new interface.
4. Fill in the postponed digital features with those digital features that require a difficult implementation effort or that saturate at three in development time.

Continuing with the case presented for designing digital behaviors for a food delivery service, the digital behavior blueprint is as follows (see Table 8.13):

Table 8.12 *Digital blueprint*

DIGITAL BEHAVIOR BLUEPRINT								
Interfaces								
Digital Conduct (description)								
Mild Implementation Effort								
Moderate Implementation Effort								
Major Implementation Effort								
Postponed Digital Features								

A table for identifying the different interfaces of the digital service and the associated digital features for each, according to their implementation difficulty.

Table 8.13 *Example of digital blueprint for the digital food ordering service*

DIGITAL BEHAVIOR BLUEPRINT

Interfaces	Enter	Selection	Order Food	Social Aspects	Reinforcing Attributes	Payment	Order Tracking	Get Food
Digital Behavior	– Sign up. – Log in. – Request password.	– View restaurants.	– Order food. – Schedule food.	– Select Shared Menus.	– Select food. – Select food from various restaurants.	– Individual payment. – Group payment.		
Mild Implementation Effort	– Register. – Log in to the app. – Password.	– Variety of different restaurants			– Filters: home-made food; foods that make you sleepy; non-staining food; food with friends; family menu; 10-minute meals; healthy food; breakfast foods; find favorite restaurant – Healthy icon – Show calories		– Follow order on map.	
Moderate Implementation Effort		– Show recommendations.			– Follow order on map.	– Promotions for total money spent. – Promotions for restaurant consumption. – Redeem for points. – Option to split the bill at the end.	– Communicate with delivery service.	

Table 8.13 *(cont.)*

			DIGITAL BEHAVIOR BLUEPRINT				
Interfaces	Enter	Selection	Order Food	Social Aspects	Reinforcing Attributes	Payment	Order Tracking	Get Food

Major Implementation Effort							
Postponed Digital Features	(1) Individuals may upload their own culinary creations to the platform, thereby allowing other users to order these dishes; (2) Users have the capability to register their favorite restaurants on the platform; (3) The system offers the functionality to schedule diets by providing accessible dietary plans; (4) The platform enables reservations at restaurants, including the availability to secure a table and the option to reserve space for another individual and their companion, complete with a prepaid ticket; (5) The service incorporates social networking features, such as profile viewing, experience sharing, and connecting with others who order similar dishes or visit the same restaurants concurrently. It includes a connectivity map indicating frequent dining connections. There is a special app promotion for adding a new contact and dining with them using the split payment option. It also features a daily caloric intake tracker; (6) The app provides "excuse menus," which simulate calls to facilitate fictitious activities.						

In this example, it can be observed how the digital behavioral map has been accommodated, in addition to incorporating new interfaces into the digital service. For the placement, it has been selected which digital features could be implemented in an initial phase, postponing those that could not for another time (postponed digital features). Moreover, for each interface, the digital behaviors that the user can perform and their reinforcers (digital features) have been associated.

It must be remembered that this digital behavior blueprint corresponds to the design of a proof of concept, so the digital features are the basic ones to start the business. In this specific example, it was detected that a communication phase with the courier was not included in the digital behavior map. This implied that in the digital behavior blueprint, we had to add a phase that was "order tracking."

8.4 Postponed Digital Features Analysis

Finally, to conclude, a special table can be created to explain the reasons why certain digital features were decided to be postponed (see Tables 8.14 and 8.15).

Table 8.14 *Postponed digital features analysis*

Digital Behavior	Reinforcers	Reason to Postpone

A table to justify why some of the digital features identified during the digital behavioral design are postponed.

Table 8.15 *Example of postponed digital features analysis*

Digital Behavior	Reinforcers	Reason to Postpone
Independent chefs can upload their menus (Dark-kitchen)	Money	The entire feature must be designed from the ground up. It is a complex subject due to the legal aspects involved. Furthermore, the technological service must be carefully managed so that it's realized and not a mere whim or a one-time occurrence. The reach of these restaurants must also be analyzed.
Register your favorite restaurant	Favorite food; discount; personalized attention	The entire feature must be designed from scratch. Here, the restaurant's own customers become ambassadors of the digital service.

Table 8.15 (*cont.*)

Digital Behavior	Reinforcers	Reason to Postpone
Follow diets	Different diet plans; health; healthy food; light food; selected menu	The entire feature must be designed from scratch. Moreover, it should be considered who will design the diets to ensure they are truly committed to. It should also be considered who will provide these menus and what benefits there would be for the restaurants.
Reserve restaurants	Getting out of the house/ office; table availability; social interactions; not cooking; eliminating thoughts of guilt (gifts)	The entire feature must be designed from scratch. There are different ways to reserve such as individual, group, or different forms of payment like individual, split payment, points …
Write and see your social network	Gourmet Journal (write experiences with dishes and receive social interactions), get to know other foodies (for sharing similar tastes or because they go to your restaurant when eating), see a map of people you usually eat with, calorie count ingested	The entire feature must be designed from scratch. Additionally, the implementation of artificial intelligence algorithms to find similar profiles should be included.
Order food that comes with the excuse of avoiding workplace meals	Avoid undesirable foods	This digital behavior is somewhat anecdotal and very attractive for media and advertising. The entire feature should be designed from scratch. Besides, it should be legally consulted what it would imply to simulate a voice. It should also explore the lack of desire for this feature, as well as not using it to avoid behaviors such as workplace dietary controls, or even cohabitation in couples.

In this example, it can be seen how it has been justified that digital features such as enabling independent chefs to upload their menus, or allowing users to register favorite restaurants among others were postponed.

This methodology for the design of digital products and services provides a framework that encapsulates all the knowledge presented through the book's chapters. However, even though a sequence of stages is offered, these can be flexible in terms of time and resource demands. For example, in the "digital behavior analysis" phase, the digital behavior designer already has sufficient information to make a basic proposal to create a quality digital product or service. The "operants design" phase represents an augmentation of the product, enabling precise design of the reinforcer. Consequently, it falls to the digital behavior designer to determine the application of this methodology and the level of flexibility they desire to employ in their work.

CHAPTER 9

Ethics in Digital Behavior Design

9.1 What Is Ethics in Technology?

Ethics in the design of digital behaviors is a matter of applying virtues rather than enforcing rules. These virtues often collide with the reality of the business model, which is the Midas king of corporations. Regardless of virtues, the sheer excitement of creating something new can overshadow certain designs that promote unethical behaviors. Even the internal dynamics of the workgroup (governed by agile methodologies) can blind us to very important aspects, such as the encouragement of risky behaviors that could have negative effects on some users. Discerning what is right and wrong in the design of digital behaviors is a challenge in the professional field of human factors, as it is not usually incorporated into the product design and development processes (Gogoll *et al.* 2021). Furthermore, considering the ethics of product design can be seen as a burden by many stakeholders, making it difficult for lower echelons within corporations to consider raising issues related to ethical concerns. A phrase attributed to Professor Amitai Etzioni from the Harvard Business School in the 1980s encapsulates this sentiment: "we teach people how to put small toys into large boxes so they seem bigger. We put hot colors onto boxes to produce impulsive buying. If you want us to teach ethical behavior, we're out of business" (Paulas 2017). It seems that the ethics of product design and development should be addressed from outside the business model itself. This is due to the problems that a worker may face if they report the risks that certain product features may pose to customers (i.e., OpenAI dismissed Sam Altman due to the risks associated with General Artificial Intelligence at the end of the year 2023, but subsequently rehired him following internal pressure from employees (Chafkin & Metz 2023)). Nevertheless, the first voices of economists and business leaders are now emerging, suggesting that there should be a corporate and political response to the need for regulating ethics in business beyond profit generation ("The New York Times" n.d.). All major technology companies have some form of code of conduct or ethics department (Google AI n.d.; Meta n.d.; Microsoft Legal n.d.), although their exact function is not entirely clear (Lauer 2021). The World Economic Forum published a White Paper in 2020 titled "Ethics by Design. An organizational approach to responsible use of technology" where they provide a manual for the implementation of ethical aspects in business (World Economic Forum 2020). In this sense, the most important point for ethical design is to remember it at the time of designing. For this to happen, the organizational culture of the company must be thoroughly permeated by the ethical principles that govern it, despite the costs to profits that may ensue in the short term.

As society's dependence on the digital services of technological tools increases, a greater ethical debate is demanded about their impact on individuals and social dynamics. Ethically reflecting on these digital services is a complex and elaborate task, as they are capable of altering the structures of social life. Thus, the more pervasive technology becomes in the daily lives of individuals, the more an ethical debate is needed not just about its use, but also about its design. In this regard, designs that may seem simple at first glance can have a significant impact on individual health; for instance, interacting with social media can affect a person's mental health (Allcott *et al.* 2020; Beyari 2023; Lee *et al.* 2022; Reed *et al.* 2023) or asking for gender in the sign-up process can reinforce stereotypes (Hentschel *et al.* 2019). Ethical deliberation of the features of a technological service is not limited to a simple financial cost-benefit analysis, but also to the value of these for the individual and society. The field of ethics involved in the creation of these design rules is normative ethics, while applied ethics (deontological and utilitarian ethics) refers to the specific uses of the technological service. Ethical deliberations about the design of the features of a technological service fall primarily on the company, but each corporate stratum must deliberate on different issues. From stakeholders to developers, through digital behavior designers and user experience designers. If any of these actors are asked if ethics matter, they would all emphatically say yes. However, economic interests often prevail over ethical principles, making it easy to fall into "ethical heuristics" that somehow validate the actions being taken. Nonetheless, not all the personal and social costs involved in some decisions in the design of these digital services can be reduced to the benefits that a company must obtain; accepting an ethical normative framework from the beginning of the design can lead to more economically efficient systems and processes in the medium and long term (van der Burg & Swierstra 2013). Adopting an ethical design from the outset in the development of these digital services could promote customer trust, improve retention rates, enhance employee behavior (trust in the company), avoid social media criticism, have a positive impact on brand recognition, increase positive reviews, and assist in negotiations for new partnerships (Chughtai *et al.* 2015; Forbes 2023). Therefore, it is advisable to avoid these "ethical heuristics" and practice real ethics in the design of digital behaviors.

9.2   When Does Ethical Deliberation Begin in the Design of Digital Behaviors?

The ethical deliberation of digital behavior designers commences from the moment the drives and reinforcers to be implemented in the digital service are proposed. As these professionals study and analyze these drives and reinforcers, the responsibility falls upon them to consider the ethical appropriateness of their transformation into digital features. For this reason, the development of a Code of Professional Ethics capable of guiding those who design and develop technological services is necessary, so that they may question their methodology and decision-making (Davis 1998). In general terms, existing Codes of Ethics and Codes of Conduct (CoC) are typically founded on three purposes: (1) to guide toward correct behaviors, (2) to serve as a benchmark for judging the ethics of behaviors carried out previously, and (3) to define the external image of the profession (Gogoll *et al.* 2021). Some of these CoCs related to software development, which could be applied to the design of digital behaviors, are those proposed by the Human Factors and Ergonomics Society (HFES n.d.), the International Labour Organization (ILO n.d.) and the Chartered Institute of Ergonomics & Human Factors (CIEHF n. d.)[1]. These CoCs mostly agree on their fundamental principles, especially highlighting the respect for human and animal dignity. Another principle worth mentioning is the necessity of using a systems approach for the design and development of digital services. The systems approach

---

[1] The design of digital behavior could be considered a sub-branch of human factors, which means that these principles would also apply to designers of digital behaviors.

proposed in this book is based on evidence from behavioral and cognitive sciences. Through this evidence, a system has been constructed that studies and proposes various methodologies to find solutions that meet the individual and collective needs of potential users of the digital service. Other principles put forth by these CoCs are those related to qualification for practice, discrimination, safety, integrity, among others. Although most CoCs agree on the main principles, they scarcely present a normative framework for guidance in the development of digital products (Gogoll *et al.* 2021). In addition to the lack of a normative framework, other barriers that could hinder the design and development of digital services according to an ethical guideline might include cognitive biases, psychological tendencies, and moral rationalizations (Schwartz 2017). However, it is advisable to explore the existence of unique principles for digital behavior designers since they apply very specific knowledge of psychology in the design of digital services. In the search for these specific ethical norms, it is important to identify the essential desirable value of the professional practice of digital behavior designers. The principle that can be used to analyze the real value contributed by the design of digital behaviors is utilitarianism, which defines value through the satisfaction of a daily life need of the user. Utilitarianism considers that a digital service provides value when it is capable of solving the need of one or more users by providing them with feelings of security, comfort, and happiness; feelings elicited by the emotion generated from the user's interaction with the virtual elements of a digital environment and the resolution of their O^D. The greatest risk that can stem from satisfying the needs of the user is the potential to create dependence on these digital services for happiness. And this poses a problem, since the compulsive and continued use of digital services can impact users' health (Andreassen *et al.* 2016). While it has not been determined that digital services are the cause of compulsive behavior, there are numerous references that point to a relationship between the use of apps and web services with some psychological distress, predominantly anxiety or depression. For these reasons, digital behavior designers must be aware of these risks and should incorporate design features that help detect and assist users who exhibit problematic behaviors in the use of their digital services. Before proposing possible features or actions to reduce the impact of these digital features on users, the problematic behaviors that may arise in the digital medium, especially online, will be analyzed.

9.3 Problematic Online Behaviors

As has been observed in previous chapters, the design of these digital spaces can lead to the formation of habits, which may become potentially addictive (Flayelle *et al.* 2023). Part of the problem lies in the traditionally minimal requirement for knowledge of behavioral and cognitive processes when designing digital products. Essentially, developers gravitate toward features that increase usage frequency, time spent using the service, or other key behaviors such as interactions or purchases (Alrobai *et al.* 2014), without realizing that often these behaviors may be indicative of pathological behaviors. The term "addiction" to define these compulsive states of use is controversial (Panova & Carbonell 2018), although in recent years it seems to be gaining momentum to be referred to as a subtype of "behavioral addiction" (to distinguish it from substance-based addiction) (Rosenberg & Feder 2014). Behavioral addictions can be defined as risk behaviors due to the attainment of short-term reinforcers that may result in persistent behaviors despite an individual's awareness and understanding of the adverse consequences (Grant *et al.* 2010; Petry 2015; Rosenberg & Feder 2014). The American Psychiatric Association (DSM-5) (American Psychiatric Association 2013) and the World Health Organization (International Classification of Diseases, Eleventh Revision [ICD-11]) (World Health Organization 2019) have recognized behavioral addiction as its own clinical entity. In relation to digital behavior, it has not yet been officially recognized as a specific disorder, although the DSM-5 and the ICD-11 are considering introducing Internet Gaming Disorder (IGD). Beyond IGD, other types of problematic online behaviors have been identified as potential digital behaviors that might meet the criteria for a behavioral addiction such as excessive online shopping (Müller *et al.* 2019), compulsive cybersexual activities (Stark *et al.* 2018), craving for social networks (Hormes *et al.* 2014), and other online apps (Fineberg *et al.* 2022). These problematic behaviors seem to share certain biological bases and cognitive, motivational, affective, and interpersonal processes (for further elaboration, see Flayelle *et al.* 2023). Although digital tools alone do not provoke neurochemical changes in our brain to cause addiction (as drugs do), the operational elements with which they have been designed may foster the acquisition of digital and online behaviors, which under certain conditions could become problematic (Hellberg *et al.* 2019).

Elements of the digital environment such as Ss can be considered a "nudge" to the user to promote certain behaviors, pursuing economic benefit for the company developing the digital service. Ethically, nudges have been extensively associated with the obstruction of positive freedom or the ability to choose freely, even evoking a certain paternalism by "helping" users choose what they need (Thaler & Sunstein 2009). However, in the design of digital behaviors, these nudges could signal the possibility of accessing reinforcers that might promote the development of digital problematic behaviors. Moreover, the fact that these problematic activities are online (instead of offline) may pose a greater risk for the development of addictive behaviors, which indicates that the very structure of the digital spaces could be a variable affecting the development of these problematic behaviors (Flayelle *et al.* 2023). It is for this reason that the design of digital behaviors should be guided by an ethical normative framework that considers this risk (Congiu & Moscati 2022). Some examples of unethical use of functional elements in the design of digital behaviors can be seen when social networks do not show all notifications in real-time, but manage them following fixed interval reinforcement schedules (after a certain time, the first time you connect, all notifications appear together), which can negatively influence individual behavior (Hormes *et al.* 2014). In video gaming, the earning of points for killing enemies follows a fixed ratio reinforcement schedule; obtaining special prizes may follow a variable ratio schedule. Promoting these types of reinforcement schedules can generate consumption habits with much resistance to extinction. Therefore, the management of these behavioral programs can become a problem when a random-ratio reinforcement schedule is used, as the level of uncertainty is maximized with the user unable to anticipate outcomes based on previous experiences (Robinson *et al.* 2019). The Triple-A model describes how the design of digital services that promote reinforcers for easy consumption of service (accessibility), at little or no cost (affordability), and that allow for anonymity (anonymity) can encourage the emergence of these problematic online behaviors (Flayelle *et al.* 2023). Table 9.1 shows the main problematic online behaviors and their relationship with the operating principles (for a more thorough review, see Flayelle *et al.* 2023).

The design of certain features in digital services may foster the emergence of problematic online behaviors, particularly among vulnerable populations. This problematic online behavior can be characterized by a loss of behavioral control and the maintenance of behavior despite negative consequences; the application of operant principles of behavior

Table 9.1 *Relationship between digital online behavior, nudges, and potentially problematic consequences*

Digital Online Behaviors	Discriminative Stimuli (nudges) & Reinforcers	Potentially Problematic Consequences
Gambling	Sign-up email bonuses; smartphone notifications on bonus bets; free bet happy hours; multi-bet offers; improved odds and cash-out offers.	Reduction in the perception of control while gambling.
Video gaming	Purchasing an item to bypass an obstacle; realistic graphics; novel rewards; surprise mechanics; player's enjoyment; social interactions.	Lack of self-control, especially in those players who previously present problems in self-regulation and decision-making.
Sexual Behavior	Pornography video or virtual reality pornography; live sex on the Internet (chats or Xcams); offline sexual partners; sexually explicit videogames; sexts; websites or apps include algorithm-based recommendations of content; personalized notifications; advanced search functionality; rating; commenting on content by other users.	They can provoke loss of control and increase the individuality as to how the inherent characteristics and traits of cybersexual activities are valued by the individuals. For example, the availability of large amounts of sexual material and its immediate access.
Shopping	Advertising pop-ups; time-limited discount offers; flexible payment options; points programs.	Online shopping expectations (variety, anonymity, immediate positive feelings) can mediate in compulsive buying behavior on the internet Trotzke *et al.* (2015).
Social Networking	Infinite scrolling; the "newsfeed" that matches a user's interests; "like" feature; "double tick" function; external cues.	Users with low self-control are especially vulnerable.
On-Demand TV streaming	Notifications about the release of new TV series; algorithm-based content recommendation; the autoplay or post-play feature; accelerated viewing; smart download; playback mode.	Significant correlations have been found between binge-watching and depression, anxiety, stress, loneliness, and insomnia Raza *et al.* (2021).

might be involved in its emergence. To some extent, digital services seem to adapt to personal motivations and exploit users' personal vulnerabilities to create forms of compulsive use (Flayelle *et al.* 2023). The administration of reinforcing stimuli could be a way to exploit certain psychological vulnerabilities in a segment of the population, thereby increasing the likelihood of inducing digital problematic behaviors. Achieving certain goals can be considerably reinforcing in a context of need, so digital spaces may become a haven for those individuals who exhibit certain cognitive limitations (e.g., low self-control, decision-making, inhibitory control, emotion dysregulation, overvaluing the importance of online rewards) and/or psychosocial vulnerabilities (e.g., stress, lack of social reinforcers ...) (Blasi *et al.* 2019; Flayelle *et al.* 2023; King *et al.* 2010). Furthermore, there seems to exist certain neural predispositions for vulnerability such as hypoconnectivity in frontostriatal networks, which may weaken goal-directed decision-making. Conversely, a hyperconnectivity in two corticostriatal pathways appears to reflect a greater regulatory control over habitual behaviors (Ersche *et al.* 2020). Therefore, the use of operant principles for the design of digital behaviors should be accompanied by an ethical normative framework that minimizes the potential risks of developing compulsive behaviors in a vulnerable segment of the user population.

9.4 What Can a Digital Behavior Designer Do?

The normative framework must be applied in actual practice. Typically, in a competitive environment such as the private sector, taking time to consider how to apply the ethical normative framework to the digital services being developed is an uncommon practice. Often, it is forgotten that design can have a negative impact on people, and moreover, the very competitive structure based on financial profits punishes any proposal that goes against reducing consumption frequency. George Herbert Mead, a social philosopher from the early twentieth century, emphasized that ethics can only confront its applicability when faced with interests, which in this case are usually corporate: "the individual should take into account all of those interests and then make out a plan of action which will rationally deal with those interests. That is the only method that ethics can bring to the individual" (Mead 1923). From the perspective of digital behavior design, adopting measures to detect those individuals who exhibit behaviors indicating certain vulnerabilities (e.g., low self-control) to the use of any digital service is an ethical imperative. The following is a series of norms for the ethical design of digital behaviors:

1. It is the duty of the digital behavior designer to question during the process of designing and developing digital services whether excessive use of the service could lead to negative consequences in the user's daily life.

2. If higher echelons explicitly urge to generate a design that causes problematic digital behaviors, stakeholders should be warned of the negative consequences for individuals and the company itself that this could entail in the medium and long term (e.g., poor corporate identity, lawsuits . . .). The digital behavior designer is advised to carefully study the emotional and legal consequences of participating in such activities.

3. Stakeholders should be cautioned about the potential compulsive use of the technological service. This warning should be accompanied by: (1) potential risks for the company (e.g., if it comes out in the media it produces a bad image for investors), (2) a proposal to mitigate these effects (actions within the service to help promote self-control strategies), and (3) the pointing out of the potential positive consequences of implementing containment actions (e.g., advertising, better brand valuation).

4. In the face of stakeholder refusal to pay attention to those users who use the service compulsively, the digital behavior designer is free to use the degree of knowledge about behavioral and cognitive sciences in the design of digital behaviors.

5. To raise user awareness of their behavior, it is recommended to implement easy-to-access usage statistics interfaces that send summaries periodically.

6. Faced with the risk that the service could result in some type of risk behaviors, the designer should propose to the team the development of algorithms capable of identifying compulsive use behaviors and alerting the individual to the type of behavior being carried out.

The subject of applied ethics within a work environment is contentious, as those who establish the norms often do not adhere to them, rendering their drafting reflective of an idealistic world. In reality, implementing ethical standards in the private sector can have unintended consequences for the individual enforcing them. It would be simplistic to suggest that if a company refuses to enforce ethical standards, the digital behavior designer should resign. Recommending such measures in real-life scenarios tends to be disconnected from reality, and this is one of the primary reasons ethical norms are not enforced. Resignation or job loss can have a significant impact on an individual's quality of life, leading them to resort to ethical heuristics to rationalize not applying ethical norms. Therefore, in everyday practice, ethical consideration in the design and development of digital services is infrequent, and we cannot fault individuals since everyone's life circumstances are unique and unknown to others. In this regard, the guidelines suggested in this chapter leave it to the designers to assess their own reality and also provide recommendations, which in some way mitigate the negative effects that the digital service might have on vulnerable individuals. The ideal scenario would be for stakeholders to be aware of the risks posed by the design of certain behavioral elements or nudges for certain populations, and to agree to integrate into the technological service certain features to prevent and minimize the harm associated with problematic digital behaviors. Listed below are a series of measures that could serve as examples of features to integrate with the technological service:

- Protocols that detect the frequency with which the same user accesses the service within a certain time frame. If the frequency of use of the digital service exceeds two or three standard deviations from the average frequency of use of the rest of the users, for several consecutive days, it could be considered that this user may have low self-control (SC). This way, they would be labeled as an at-risk user.
- When atypical behavior is detected, as proposed in the previous case, a notification should appear to alert the user of their conduct. This notification could contain a short and concise text about the risk behavior: "You have disproportionately exceeded the number of times you access the service compared to other users. This type of behavior is risky and can lead to technological addiction. To continue using the service you must read the following document containing information on the negative consequences of your behavior."
- The document should contain a list of negative consequences (e.g., suffering depression due to decreased social contact, anxiety if usage expectations are not met, deterioration of family relationships, social relationships, academic/work performance, low life satisfaction, etc.).
- To demonstrate potential self-control strategies to reduce the propensity for disproportionate use of the digital service. For instance, proactive strategies (top-down behavioral regulation) and reactive strategies (bottom-up behavioral regulation) can be proposed (McRae

et al. 2012). Proactive measures could include actions such as leaving the cellphone in inaccessible places, turning off notifications, temporary deletion of the digital service, scheduled usage times, etc. On the other hand, reactive strategies involve being aware of when one wishes to access the digital service, engaging in incompatible tasks such as reading a book, or self-verbalizing the positive consequences of reduced use of the digital service such as time to spend with family and friends (Brevers & Turel 2019).

- A feature that allows the user to self-block the digital service, thus regulating a schedule/time of use. This feature to block the app could also be given to another person who would externally regulate the schedule/time of use.
- A usage statistics section within the digital service, where the user can see how their usage compares with that of other users. Ideally, there should be a clear marker of the user's level of compulsion.
- Signing a responsibility of use agreement would also be advisable in cases where risky behaviors are detected and to regain access to the technological service.
- Providing the user with external resources to become informed about the risks of technological behavioral addiction and how to address them.

The following presents a diagram of the potential user pathway upon detection of a risk behavior when utilizing a digital service (see Figure 9.1).

Figure 9.1 depicts a possible flowchart for managing the detection of behavioral patterns that may indicate the presence of problematic digital behaviors. Initially, an algorithm specific to the digital service must be integrated to detect digital behavioral parameters of risk behaviors, such as accessing the app 2 or 3 standard deviations above the mean of other users. These parameters should be present for a set duration of days or weeks to distinguish them from isolated anxiety-inducing situations. Once a stable behavioral pattern is identified that signals the potential existence of a problematic digital behavior, a notification is sent to the user alerting them of an irregular pattern in their usage, which may be related to certain behavioral addictions. Subsequently, a document is presented describing the clinical signs and/or symptomatology of these behaviors, followed by a checklist to help users recognize any of these signs and/or symptoms. The purpose of this checklist is to raise the individual's awareness of their own condition. To ensure that the user reads the options and does not simply click "next," they may be required to respond to some of the

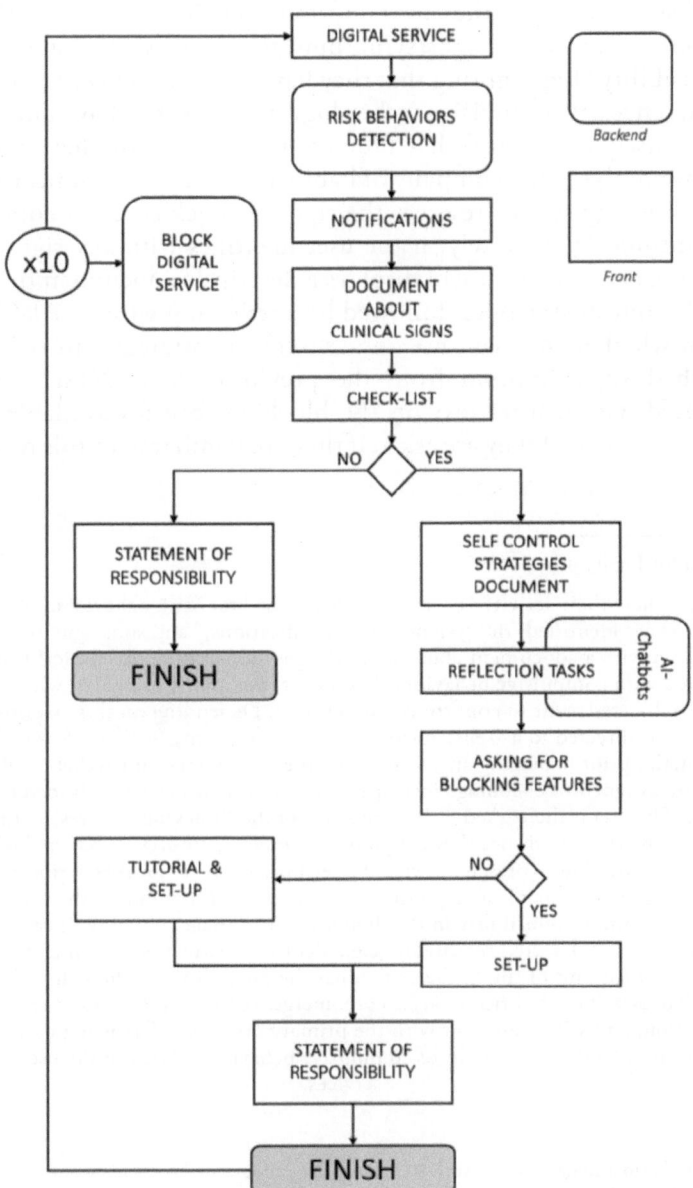

Figure 9.1 Flowchart for the management of behavioral pattern detection indicative of problematic digital behaviors in digital service users

The diagram depicts an algorithm designed to identify and manage compulsive behaviors in digital service users. It begins with the user's interaction in a "digital checklist items before the "next" option is enabled.[2] Then, if the user does not recognize any signs/symptoms, they must sign a "statement of responsibility" highlighting that they have been advised of the negative consequences of compulsive technology use, that they have understood these consequences, and that they are responsible for their behavior, thereby absolving the company of liability[3]; and a clause indicating that the technology service reserves the right to block access if compulsive use continues. Conversely, if the user identifies with any signs/symptoms, the activated protocol displays a descriptive document on different self-control strategies, followed by a reflection with an AI-Chatbot bot on whether they will use any self-control strategies to reduce the identified sign/symptom from the previous phase. Next, users are presented with a brief text on the blocking features available in the digital service, and they are asked if they are familiar with this feature. If

------

Caption for Figure 9.1   (cont.)

service," where their activity is continuously assessed for "risk behaviors." When such behaviors are identified, the system sends "notifications," and subsequently provides the user with a "document about clinical signs" illustrating the clinical features associated with compulsive behaviors. Thereafter, the user is presented with a "check-list" for self-assessment to confirm certain criteria. Depending on this assessment, the user may be directed to a "self-control strategies document," which offers techniques and methods for managing and controlling their behaviors. In parallel, "reflection tasks" are assigned to encourage introspection about their behavior, guided by an AI-chatbot. The user is then asked if they are aware of the "blocking features" of the digital service. If the response is no, they are guided through a "tutorial & set-up," whereas if yes, they are simply presented with the set-up screen. In both scenarios, before concluding, a "statement of responsibility" is presented, reminding the user of their autonomy and responsibility in the digital service's usage. Notably, if certain risk behaviors are detected repeatedly (e.g., algorithm activation ten times within a predetermined unit of time), the system has the capability to "block digital services" for that user. This algorithm, at its core, merges digital intervention techniques, education, and self-assessment with the primary objective of assisting users in the identification and management of compulsive behaviors related to the use of digital services.

------

[2] It is possible to include an option such as "I do not recognize any of these signs/symptoms," or simply include in the checklist statement that if you do not recognize any of the signs/symptoms, mark the one you would be most afraid of developing. It is important to carefully consider these statements, as sometimes they can lead to over-interpretation of symptoms and "induce" a disorder when there is none (Foulkes & Andrews 2023).

[3] This type of clause exempting the company from any liability is usually liked by stakeholders, which can be a reason for them to authorize and finance these measures.

the answer is no, they are offered a tutorial and setup; if yes, only the setup is offered. Finally, on both paths, the statement of responsibility is presented, and once read and approved, the user is returned to the digital service. An interesting option to implement is that if risk behavior is detected a specific number of times within a set period (fixed interval reinforcement schedule with a duration requirement; in Figure 9.1 this is shown as "x10"), the digital service could be blocked. This measure should be considered with stakeholders and an ethics committee if possible.

In conclusion, it is essential for designers of digital behaviors to incorporate an ethical space in their practices to reflect on the impact of applying operant and cognitive principles in the design and development of digital services. This need arises due to the potentially harmful effect such actions can have on vulnerable populations. This ethical space must be acknowledged and valued by stakeholders as of paramount importance, ensuring that ethical decisions are adopted uniformly, from the highest corporate levels to the design and development stages. As professionals in the field of Human Factors and Engineering Psychology, it is recommended to care for the health of users who interact with our designs. This principle is supported by numerous Codes of Conduct (CoC), including the Ethical Code of the American Psychological Association. This code highlights general principles such as Beneficence and Nonmaleficence, Fidelity and Responsibility, Integrity, Justice, and Respect for People's Rights and Dignity (American Psychological Association 2017). Therefore, it is the duty of digital behavior designers to ensure that these principles are reflected in all aspects of their work.

Glossary

Allostasis: Allostasis is defined as a brain process that allows the maintenance of physiological stability of the body while producing functional changes that enable us to actively adjust to predictable or unpredictable events.

Antecedent: Stimuli or previous situations that can influence the response or behavior of a user toward a digital service.

Anticipate: To anticipate refers to the ability to foresee future needs or events based on previous experiences or the observation of environmental patterns and regularities.

Appetitive phase: This phase involves a set of learning-sensitive behaviors for exploring the environment with the goal of obtaining the reinforcer that satisfies their drive.

Appetitive stimulus: Are those stimuli that produce a pleasant hedonic response in the organism.

Aversive stimulus: Are those stimuli that produce an unpleasant hedonic response in the organism.

Behavior: Interaction of an individual in a physical environment.

Behavioral competition: Behavioral competition is conceptualized as the constant struggle for the user's time among various alternative behaviors.

Behavioral oscillation: Variations in user behavior in response to different stimuli or in different contexts.

Behavioral profiling: It is used to identify the most appropriate period for a user to utilize a service, evaluating the time in days, weeks, months, or even years.

Behaviorism: A theory that maintains that the behavior of humans and animals can be explained by their learning history instead of internal representations of possible futures.

Cognitive unit: A mental representation that the user utilizes to make anticipations of consequences and control voluntary behavior.

Cognitivism: Cognitivism and its 4Es explain that behavior is the result of interactions between our cognitive system and the environment, but mediated by mental representations (or models of the world) that are formed as a result of previous interactions.

Consequent: Consequent or contingent stimuli refer to those that increase the probability of a behavior being repeated (reinforcers) or decrease that probability (punishments).

Consummatory phase: Final behavior in which the organism consumes the reinforcer.

Contiguity: Is the temporal space between the occurrence of two stimuli or between the issuance of a response and its consequent stimuli.

Contingency: Contingencies are the probabilistic relationships between behaviors and their consequences.

Contingent or consequent stimulus: Stimuli that are administered in relation to a specific behavior.

Delayed reinforcer: A reinforcer whose administration does not occur immediately after the desired behavior, but after a lapse of time, which can range from several seconds to days.

Direct reinforcer: Reinforcers offered directly by the technological service.

Digital behavior: Interaction of an individual in a digital environment.

Digital reinforcement schedule: Strategies used in operant conditioning to increase the probability of digital responses through the administration of reinforcers.

Discriminative stimulus: A stimulus that signals the availability of a reinforcer or punishment in the presence of a specific response.

Discriminatory function: Effect of the discriminative stimulus on the appropriate behavior to obtain a reinforcer.

Digital operant box: It is a psychological construct that alludes to and theoretically recognizes Skinner's Box, with the objective of recreating the learning principles derived from the experimental analysis of behavior and finding their analogs in the digital and virtual environment.

Drive: An internal vector that signals the valence and arousal of the behavior.

Encoding: The process through which past experiences are converted into memories stored in neural clusters known as engrams.

Feedback function: The effect exerted by the consumption of a reinforcer on the organism.

Goal-directed behavior: Behavior resulting from the interaction between a specific internal state, the anticipation of the value of the reinforcer, and the initiation of a sequence of behaviors aimed at seeking, obtaining, and consuming the reinforcer to satisfy an internal state.

Goal-directed control behavior: Regulation of voluntary behavior through goal setting.

Goal–value representations: Representation of the expected value or utility of a reinforcing stimulus that is contingent on the selection of a particular behavior.

Human factors and psychology engineering: A field of study focusing on human interactions with machines and how to improve their interactivity.

Indirect reinforcer: Reinforcers that activate a system or process through which the reinforcer is obtained outside of the digital medium, in the physical reality.

Immediate reinforcer: Technological service reinforcers obtained by users with low contiguity between the behavior and the reinforcer, meaning the time elapsed between the digital behavior and the appearance of the reinforcer is zero or a few seconds.

Interaction: Action-reaction response between two or more autonomous units.

Long-term memory: The capacity to evoke memories without a physical precept that were encoded and stored in the past.

Matching: Mathematical relationship between the time spent interacting with a digital object and the proportion of reinforcers obtained from this behavior.

Mental representation: This refers to a neuronal representation that is activated in working memory, with which the subject can manipulate and of which they are subjectively aware.

Molar behavior: Molar behavior in the design of digital behaviors is intentionally directed toward goals and is composed of smaller segments of behaviors (molecular) that aggregate to form the molar behavior.

Molar reinforcer: A principal reinforcer with the power to satisfy the drive with which a digital product or service is used. It is related to molar behavior and is obtained in the consummatory phase of the behavior.

Molecular behavior: Smaller segments of behaviors that aggregate and ultimately shape the resulting molar behavior.

Molecular reinforcer: Reinforcers that guide the user's behavior through a digital product or service until obtaining the molar reinforcer. It is related to molecular behavior and is obtained in the appetitive phase of the behavior.

Motivated behavior: This is the interaction between a determined internal state, the anticipation of the reinforcer's value, and the initiation of a series of motor actions with the goal of consuming the reinforcer to satisfy the internal state.

Neuronal representation: It is the configuration of an environmental or internal stimulus within a neuronal space created from the encoding of this object.

Operant response: A functional relationship that maintains the voluntary behavior of an individual with the sensory elements of their environment (stimuli).

Prediction: The ability of a system to possess prior knowledge about the cause-and-effect relationship of an event.

Punishment: Punishments refer to stimuli administered in relation to a specific behavior that decrease the likelihood of that behavior being repeated.

Recycling neural hypothesis: A theory on how the use of cultural tools transforms the structure and connectivity of neural groups.

Reinforcement: A process to increase the performance of certain behaviors through the administration of reinforcers.

Reinforcement schedule: Strategies used in operant conditioning to increase the likelihood of a response through the administration of reinforcers.

Reinforcer: An appetitive stimulus with sufficient power to increase the probability of a behavior's occurrence.

Reinforcer matrix: A set of potentially obtainable reinforcers for a person by performing a specific behavior.

Response class: It is defined as a set of response topographies that serve the same function, that is, lead to the same reinforcer.

Shaping: The process of gradually guiding the user's behavior toward a specific goal through positive reinforcement of successive approximations to the desired behavior.

Stimulus: Any object or event that affects an organism's behavior.

Users: Open learning systems of organic and computational basis capable of anticipating future events through the processing of contextual information associated with past experiences in such a way that can generate future representational expectations that will control their behavior in a manner that subjectively maximizes their benefit by resolving physiological and psychological needs through digital tools. The discrepancy between expectation and reality is what would be considered learning and would modify the neuronal representations in which past experiences are encoded.

User needs: Internal states of the user that drive goal-directed behavior.

Virtual behavior: Although it can be interchangeable with digital behavior in different contexts, here it is used to refer to behaviors carried out in virtual environments through avatars.

Working memory: The capacity to manipulate and maintain information in real time.

References

A Free Market Manifesto That Changed the World, Reconsidered. *The New York Times*. (n.d.). Retrieved November 6, 2023, from www.nytimes.com/2020/09/11/business/dealbook/milton-friedman-doctrine-social-responsibility-of-business.html.

Adolphs, R., Tranel, D., & Buchanan, T. W. (2005). Amygdala Damage Impairs Emotional Memory for Gist But Not Details of Complex Stimuli. *Nature Neuroscience*, 8(4), 512–518.

Ahn, S. J. G., Johnsen, K., & Ball, C. (2019). Points-Based Reward Systems in Gamification Impact Children's Physical Activity Strategies and Psychological Needs. *Health Education & Behavior: The Official Publication of the Society for Public Health Education*, 46(3), 417–425.

Ait Oumeziane, B., Schryer-Praga, J., & Foti, D. (2017). "Why Don't They 'Like' Me More?": Comparing the Time Courses of Social and Monetary Reward Processing. *Neuropsychologia*, 107, 48–59.

Allcott, H., Braghieri, L., Eichmeyer, S., *et al.* (2020). The Welfare Effects of Social Media. *American Economic Review*, 110(3), 629–676.

Allport, G. W. (1937). The Functional Autonomy of Motives. *The American Journal of Psychology*, 50, 141–147.

Alrobai, A., Phalp, K., & Ali, R. (2014). Digital Addiction: A Requirements Engineering Perspective. *Lecture Notes in Computer Science (Including Subseries Lecture Notes in Artificial Intelligence and Lecture Notes in Bioinformatics)*, 8396 **LNCS**, 112–118.

Alshobaili, F. A., & AlYousefi, N. A. (2019). The Effect of Smartphone Usage at Bedtime on Sleep Quality among Saudi Non-medical Staff at King Saud University Medical City. *Journal of Family Medicine and Primary Care*, 8(6), 1953–1957.

Amalric, M., & Dehaene, S. (2016). Origins of the Brain Networks for Advanced Mathematics in Expert Mathematicians. *Proceedings of the National Academy of Sciences of the United States of America*, 113(18), 4909–4917.

American Psychiatric Association. (2013). *Diagnostic and Statistical Manual of Mental Disorders 5th ed.*, Arlington, VA: American Psychiatric Publishing.

American Psychological Association. (2017). Ethical Principles of Psychologists and Code of Conduct. Retrieved June 27, 2023, from www.apa.org/ethics/code.

Anderson, C., John, O. P., Keltner, D., & Kring, A. M. (2001). Who Attains Social Status? Effects of Personality and Physical Attractiveness in Social Groups. *Journal of Personality and Social Psychology*, 81(1), 116–132.

Anderson, C., Srivastava, S., Beer, J. S., Spataro, S. E., & Chatman, J. A. (2006). Knowing Your Place: Self-Perceptions of Status in Face-to-Face Groups. *Journal of Personality and Social Psychology*, 91(6), 1094–1110.

Anderson, D. J. (2016). Circuit Modules Linking Internal States and Social Behaviour in Flies and Mice. *Nature Reviews. Neuroscience*, 17(11), 692–704.

Anderson, J. (1980). Concepts, Propositions, and Schemata: What Are the Cognitive Units? *Nebraska Symposium on Motivation* (28), 121–162.

Anderson, J. R. (1983a). A Spreading Activation Theory of Memory. *Journal of Verbal Learning and Verbal Behavior*, 22(3), 261–295.

Andreassen, C. S., Billieux, J., Griffiths, M. D., *et al.* (2016). The Relationship Between Addictive Use of Social Media and Video Games and Symptoms of Psychiatric Disorders: A Large-Scale Cross-Sectional Study. *Psychology of Addictive Behaviors: Journal of the Society of Psychologists in Addictive Behaviors*, 30(2), 252–262.

Ang, Y.-S., Manohar, S., Plant, O., *et al.* (2018). Dopamine Modulates Option Generation for Behavior. *Current Biology*, 28(10), 1561–1569.e3.

APA. (2014). A Career in Human Factors and Engineering Psychology. Retrieved June 4, 2023, from www.apa.org/education-career/guide/subfields/human-factors/education-training.

Apicella, C. L., & Silk, J. B. (2019). The Evolution of Human Cooperation. *Current Biology: CB*, 29(11), R447–R450.

Asher, S., & Cole, J. (1990). *Peer Rejection in Childhood*, Cambridge: Cambridge University Press.

Ashokkumar, A., Talaifar, S., Fraser, W. T., *et al.* (2020). Censoring Political Opposition Online: Who Does It and Why. *Journal of Experimental Social Psychology*, 91, 104031.

Astle, D. E., Johnson, M. H., & Akarca, D. (2023). Toward Computational Neuroconstructivism: A Framework for Developmental Systems Neuroscience. *Trends in Cognitive Sciences*, 27(8), 726–744. doi:10.1016/J.TICS.2023.04.009.

Aydin Gokgoz, Z., Ataman, M. B., & van Bruggen, G. H. (2021). There's an App for That! Understanding the Drivers of Mobile Application Downloads. *Journal of Business Research*, 123, 423–437.

Bacon, F. (1994). *Novum Organum*. Peru, Illinois: Open Court Publishing Company. https://archive.org/details/novumorganumooooobaco_t7v4/page/n5/mode/2up

Baerends, G. (1976). On Drive, Conflict and Instinct, and the Functional Organization of Behavior. *Progress in Brain Research*, 45, 425–447.

Bailey, M. R., Jensen, G., Taylor, K., *et al.* (2015). A Novel Strategy for Dissecting Goal-Directed Action and Arousal Components of Motivated Behavior with a Progressive Hold-Down Task. *Behavioral Neuroscience*, 129(3), 269–280.

Baker, F., Johnson, M. W., & Bickel, W. K. (2003). Delay Discounting in Current and Never-Before Cigarette Smokers: Similarities and Differences across Commodity, Sign, and Magnitude. *Journal of Abnormal Psychology*, 112 (3), 382–392.

Baldwin, B. (1913). John Locke's Contributions to Education. *The Sewanee Review*, 21(2), 177–187.

Barlow, D. H. (2000). Unraveling the Mysteries of Anxiety and Its Disorders from the Perspective of Emotion Theory. *The American Psychologist*, 55(11), 1247–1263.

Barlow, G. (1968). Ethological Units of Behaviour. In D Ingle (ed.), *Central Nervous Systems and Fish Behavior*, Chicago: University of Chicago Press, pp. 217–232.

Barr, N., Pennycook, G., Stolz, J. A., & Fugelsang, J. A. (2015). The Brain in Your Pocket: Evidence That Smartphones Are Used to Supplant Thinking. *Computers in Human Behavior*, 48, 473–480.

Barrett, R. L. C., Dawson, M., Dyrby, T. B., *et al.* (2020). Differences in Frontal Network Anatomy Across Primate Species. *The Journal of Neuroscience: The Official Journal of the Society for Neuroscience*, 40(10), 2094–2107.

Barton, K. R., Yazdani, Y., Ayer, N., *et al.* (2017). The Effect of Losses Disguised As Wins and Near Misses in Electronic Gaming Machines: A Systematic Review. *Journal of Gambling Studies*, 33(4), 1241–1260.

Basten, U., Biele, G., Heekeren, H. R., & Fiebach, C. J. (2010). How the Brain Integrates Costs and Benefits During Decision Making. *Proceedings of the National Academy of Sciences*, 107(50), 21767–21772.

Baum, W. M. (2004). Molar and Molecular Views of Choice. *Behavioural Processes*, 66(3), 349–359.

Bauman, Z. (2000). *Liquid Modernity*, Cambridge: Polity Press. Retrieved from www.wiley.com/en-us/Liquid+Modernity-p–9780745624099.

Bayne, T., Brainard, D., Byrne, R. W., *et al.* (2019). What Is Cognition? *Current Biology: CB*, 29(13), R608–R615.

Bechara, A., & Damasio, A. R. (2005). The Somatic Marker Hypothesis: A Neural Theory of Economic Decision. *Games and Economic Behavior*, 52(2), 336–372.

Bechterev, V. (1932). *General Principles of Human Reflexology*, New York: International Publishers.

Beer, B., & Trumble, G. (2014). Timing Behavior As a Function of Amount of Reinforcement. *Psychonomic Science*, 2(1), 71–72 (1965).

Belin, D., Jonkman, S., Dickinson, A., Robbins, T. W., & Everitt, B. J. (2009). Parallel and Interactive Learning Processes within the Basal Ganglia: Relevance for the Understanding of Addiction. *Behavioural Brain Research*, 199(1), 89–102.

Bellman, S., Potter, R. F., Treleaven-Hassard, S., Robinson, J. A., & Varan, D. (2011). The Effectiveness of Branded Mobile Phone Apps: https://Doi.Org/10.1016/j.Intmar.2011.06.001, 25(4), 191–200.

Bench, S. W., & Lench, H. C. (2013). On the Function of Boredom. *Behavioral Sciences (Basel, Switzerland)*, 3(3), 459–472.

Bender, A., & Beller, S. (2019). The Cultural Fabric of Human Causal Cognition. *Perspectives on Psychological Science: A Journal of the Association for Psychological Science*, 14(6), 922–940.

Ben-Naim, A. (2020). Entropy and Time. *Entropy*, 22(4). doi:10.3390/E22040430

Berger, J., Rosenholtz, S. J., & Zelditch, M. Jr. (2003). Status Organizing Processes. https://doi.Org/10.1146/Annurev.so.06.080180.002403, 6(1), 479–508.

Berger, S., & Rejman, K. (2019). Food Digestion in Ivan Petrovich Pavlov Studies on 115 Anniversary of His Nobel Prize and Present Avenues. *Roczniki Panstwowego Zakladu Higieny*, 70(1), 97–102.

Berman, D. E., & Dudai, Y. (2001). Memory Extinction, Learning Anew, and Learning the New: Dissociations in the Molecular Machinery of Learning in Cortex. *Science*, 291(5512), 2417–2419.

Bernoulli, D. (2011). Exposition of a New Theory on the Measurement of Risk. *Econometrica*, 22(1), 23–36.

Berridge, K. C. (2009). Wanting and Liking: Observations from the Neuroscience and Psychology Laboratory. *Inquiry*, 52(4), 378–398.

Berridge, K. C., Ho, C. Y., Richard, J. M., & Difeliceantonio, A. G. (2010). The Tempted Brain Eats: Pleasure and Desire Circuits in Obesity and Eating Disorders. *Brain Research*, 1350, 43–64.

Beyari, H. (2023). The Relationship between Social Media and the Increase in Mental Health Problems. *International Journal of Environmental Research and Public Health*, 20(3). doi:10.3390/ijerph20032383

Binder, J. R., Desai, R. H., Graves, W. W., & Conant, L. L. (2009). Where Is the Semantic System? A Critical Review and Meta-Analysis of 120 Functional Neuroimaging Studies. *Cerebral Cortex*, 19(12), 2767–2796.

Black, R. E., Walters, G. C., & Webster, C. D. (1972). Fixed-Interval Limited-Hold Avoidance with and without Signalled Reinforcement. *Journal of the Experimental Analysis of Behavior*, 17(1), 75.

Blader, S. L., & Chen, Y.-R. (2014). What's in a Name? Status, Power, and Other Forms of Social Hierarchy. In *The Psychology of Social Status*, New York, NY: Springer New York, pp. 71–95.

Blasi, M. D. I., Giardina, A., Giordano, C., et al. (2019). Problematic Video Game Use As an Emotional Coping Strategy: Evidence from a Sample of MMORPG Gamers. *Journal of Behavioral Addictions*, 8(1), 25–34.

Blom, T., Feuerriegel, D., Johnson, P., Bode, S., & Hogendoorn, H. (2020). Predictions Drive Neural Representations of Visual Events Ahead of Incoming Sensory Information. *Proceedings of the National Academy of Sciences of the United States of America*, 117(13), 7510–7515.

Board, S. (2009). *Utility Maximization Problem*, Department of Economics, UCLA.

Bouton, M. E. (2000). A Learning Theory Perspective on Lapse, Relapse, and the Maintenance of Behavior Change. *Health Psychology: Official Journal of the Division of Health Psychology, American Psychological Association*, 19(1S), 57–63.

Bouton, M. E. (2021). Context, Attention, and the Switch between Habit and Goal-Direction in Behavior. *Learning & Behavior*, 49(4), 349–362.

Boyd, D. E., Kannan, P. K., & Slotegraaf, R. J. (2019). Branded Apps and Their Impact on Firm Value: A Design Perspective. *Journal of Marketing Research*, 56(1), 76–88.

Brady, W. J., McLoughlin, K., Doan, T. N., & Crockett, M. J. (2021). How Social Learning Amplifies Moral Outrage Expression in Online Social Networks. *Science Advances*, 7(33). doi:10.1126/sciadv.abe5641

Brenner, S. L., Jones, J. P., Rutanen-Whaley, R. H., Parker, W., Flinn, M. v., & Muehlenbein, M. P. (2015). Evolutionary Mismatch and Chronic Psychological Stress. *Journal of Evolutionary Medicine*, 3(1), 1–11.

Brevers, D., & Turel, O. (2019). Strategies for Self-Controlling Social Media Use: Classification and Role in Preventing Social Media Addiction Symptoms. *Journal of Behavioral Addictions*, 8(3), 554–563.

Broberger, C. (2005). Brain Regulation of Food Intake and Appetite: Molecules and Networks. *Journal of Internal Medicine*, 258(4), 301–327.

Bunge, M. (1983). The Nature of Applied Science and Technology. *Philosophy and Culture*, 2. Retrieved from https://philpapers.org/rec/BUNTNO-2

Burke, P. J., & Stets, J. E. (1999). Trust and Commitment through Self-Verification. *Social Psychology Quarterly*, 62(4), 347–366.

Business of Apps. (2023). Candy Crush Revenue and Usage Statistics. www.businessofapps.com/data/candy-crush-statistics/

Byrne, T., & Sarno, B. (2019). Response Duration Is Sensitive to Both Immediate and Delayed Reinforcement. *Journal of the Experimental Analysis of Behavior*, 111 (1), 94–115.

Calabrese, J. R., Goetschius, L. G., Murray, L., et al. (2022). Mapping Frontostriatal White Matter Tracts and Their Association with Reward-Related Ventral Striatum Activation in Adolescence. *Brain Research*, 1780. doi:10.1016/J.BRAINRES.2022.147803

Calvillo, D. P., & Hawkins, W. C. (2016). Animate Objects Are Detected More Frequently than Inanimate Objects in Inattentional Blindness Tasks Independently of Threat. *The Journal of General Psychology*, 143 (2), 101–115.

Canhoto, A., & Backhouse, J. (2008). General Description of the Process of Behavioural Profiling. In M. Hildebrandt and S. Gutwirth (Eds.), *Profiling the European Citizen: Cross-Disciplinary Perspectives*, Dordrecht: Springer, 47–63.

Cannon, W. (1932). *The Wisdom of the Body*, New York: W. W. Norton and Co.

Cantlon, J. F., Brannon, E. M., Carter, E. J., & Pelphrey, K. A. (2006). Functional Imaging of Numerical Processing in Adults and 4-y-Old Children. *PLoS Biology*, 4(5), 844–854.

Cantlon, J. F., & Li, R. (2013). Neural Activity During Natural Viewing of Sesame Street Statistically Predicts Test Scores in Early Childhood. *PLoS Biology*, 11(1). doi:10.1371/JOURNAL.PBIO.1001462

Cao, L., Liu, X., & Cao, W. (2018). The Effects of Search-Related and Purchase-Related Mobile App Additions on Retailers' Shareholder Wealth: The Roles of Firm Size, Product Category, and Customer Segment. *Journal of Retailing*, 94 (4), 343–351.

Carney, J. (2020). Thinking avant la lettre: A Review of 4E Cognition. *Evolutionary Studies in Imaginative Culture*, 4(1), 77.

Carroll, M. E., Anker, J. J., & Perry, J. L. (2009). Modeling risk factors for nicotine and other drug abuse in the preclinical laboratory. *Drug Alcohol Depend* [online serial], 104(1). Accessed at: https://pubmed.ncbi.nlm.nih.gov/19136222/.

Castel, A. D. (2007). The Adaptive and Strategic Use of Memory By Older Adults: Evaluative Processing and Value-Directed Remembering. *Psychology of Learning and Motivation - Advances in Research and Theory*, 48, 225–270.

Chafkin, M., & Metz, R. (2023). What We Know So Far About Why OpenAI Fired Sam Altman. *TIME*. https://time.com/6337437/sam-altman-openai-fired-why-microsoft-musk/

Chiao, J. Y., Bordeaux, A. R., & Ambady, N. (2004). Mental Representations of Social Status. *Cognition*, 93(2), B49–57.

Chiao, J. Y., Harada, T., Oby, E. R., Li, Z., Parrish, T., & Bridge, D. J. (2009). Neural Representations of Social Status Hierarchy in Human Inferior Parietal Cortex. *Neuropsychologia*, 47(2), 354–363.

Chudek, M., & Henrich, J. (2011). Culture-Gene Coevolution, Norm-Psychology and the Emergence of Human Prosociality. *Trends in Cognitive Sciences*, 15(5), 218–226.

Chughtai, A., Byrne, M., & Flood, B. (2015). Linking Ethical Leadership to Employee Well-Being: The Role of Trust in Supervisor. *Journal of Business Ethics*, 128(3), 653–663.

Chun, J. W., Choi, J., Cho, H., et al. (2018). Role of Frontostriatal Connectivity in Adolescents with Excessive Smartphone Use. *Frontiers in Psychiatry*, 9(SEP), 437.

Chung, S.-H., & Herrnstein, R. J. (1967). Choice and Delay of Reinforcement. *Journal of the Experimental Analysis of Behavior*, 10(1), 67–74.

Churchland, P. S., & Winkielman, P. (2012). Modulating Social Behavior with Oxytocin: How Does It Work? What Does It Mean? *Hormones and Behavior*, 61 (3), 392–399.

CIEHF. (n.d.). Code of Conduct. Retrieved June 27, 2023, from https://ergonomics.org.uk/advice/code-of-conduct.html

Clutton-Brock, T. H., & Harvey, P. H. (1980). Primates, Brains and Ecology. *Journal of Zoology*, 190(3), 309–323.

Colagè, I., & d'Errico, F. (2020). Culture: The Driving Force of Human Cognition. *Topics in Cognitive Science*, 12(2), 654–672.

Colley, C. (1902). *Human Nature and the Social Order*, New York: C. Scribner's sons.

Collins, A. M., & Loftus, E. F. (1975). A Spreading-Activation Theory of Semantic Processing. *Psychological Review*, 82(6), 407–428.

Colwill, R. M., & Rescorla, R. A. (1990). Effect of Reinforcer Devaluation on Discriminative Control of Instrumental Behavior. *Journal of Experimental Psychology. Animal Behavior Processes*, 16(1), 40–47.

Congiu, L., & Moscati, I. (2022). A Review of Nudges: Definitions, Justifications, Effectiveness. *Journal of Economic Surveys*, 36(1), 188–213.

Constantinescu, A. O., O'Reilly, J. X., & Behrens, T. E. J. (2016). Organizing Conceptual Knowledge in Humans with a Gridlike Code. *Science*, 352(6292), 1464–1468.

Corbett, S., Courtiol, A., Lummaa, V., Moorad, J., & Stearns, S. (2018). The Transition to Modernity and Chronic Disease: Mismatch and Natural Selection. *Nature Reviews Genetics*, 19(7), 419–430.

Cortina, J. M., Köhler, T., Keeler, K. R., & Pugh, S. D. (2022). Situation Strength As a Basis for Interactions in Psychological Models. *Psychological Methods*, 27 (2). doi:10.1037/MET0000372

Crosscut. (2020). "Free" Casino Apps Prey on Addiction, Users Say, and WA Lawmakers Are Considering a Crackdown | Crosscut. Retrieved June 28, 2023, from https://crosscut.com/2020/02/free-casino-apps-prey-addiction-users-say-and-wa-lawmakers-are-considering-crackdown

Croxson, P. L., Walton, M. E., O'Reilly, J. X., Behrens, T. E. J., & Rushworth, M. F. S. (2009). Effort-Based Cost-Benefit Valuation and the Human Brain. *The Journal of Neuroscience: The Official Journal of the Society for Neuroscience*, 29(14), 4531–4541.

Darejeh, A., & Salim, S. S. (2016). Gamification Solutions to Enhance Software User Engagement – A Systematic Review. *International Journal of Human-Computer Interaction*, 32(8), 613–642. https://doi.org/10.1080/10447318.2016.1183330

David, L., Vassena, E., & Bijleveld, E. (2024). The Aversiveness of Mental Effort: A Meta-Analysis. *PsyArXiv*. doi:10.31234/osf.io/m8zf6.

Davis, M. (1998). Thinking Like an Engineer: Studies in the Ethics of a Profession. Retrieved from https://philpapers.org/rec/DAVTLA-3

Davison, M., & McCarthy, D. (2016). *The Matching Law: A Research Review*. London: Routledge. doi:10.4324/9781315638911

Daw, N. D., Niv, Y., & Dayan, P. (2005). Uncertainty-Based Competition between Prefrontal and Dorsolateral Striatal Systems for Behavioral Control. *Nature Neuroscience*, 8(12), 1704–1711.

Dayan, P., & Niv, Y. (2008). Reinforcement Learning: The Good, the Bad and the Ugly. *Current Opinion in Neurobiology*, 18(2), 185–196.

de Wit, S., Ostlund, S. B., Balleine, B. W., & Dickinson, A. (2009). Resolution of Conflict between Goal-Directed Actions: Outcome Encoding and Neural Control Processes. *Journal of Experimental Psychology. Animal Behavior Processes*, 35(3), 382–393.

Deci, E. L. (1971). Effects of Externally Mediated Rewards on Intrinsic Motivation. *Journal of Personality and Social Psychology*, 18(1), 105–115.

Declerck, C. H., & Boone, C. (2018). The Neuroeconomics of Cooperation. *Nature Human Behaviour*, 2(7), 438–440.

Dehaene, S. (2005). Evolution of Human Cortical Circuits for Reading and Arithmetics: The "Neuronal Recycling" Hypothesis. In *From Monkey Brain to Human Brain*, Cambridge, Mass: MIT Press, pp. 133–157.

Dehaene, S. (2009). *Reading in the Brain: The New Science of How We Read*, New York: Penguin Books.

Dehaene, S. (2014a). *Consciousness and the Brain*, New York: Penguin Books.

Dehaene, S. (2014b). *The Signature of a Conscious Thought*, New York: Penguin Books.

Dehaene, S., & Cohen, L. (2007). Cultural Recycling of Cortical Maps. *Neuron*, 56(2), 384–398.

Davis, U. C. (2016). Memory Replay Prioritizes High-Reward Experiences. Retrieved June 29, 2023, from https://biology.ucdavis.edu/news/memory-repl ay-prioritizes-high-reward-experiences

d'Errico, F., & Colagè, I. (2018). Cultural Exaptation and Cultural Neural Reuse: A Mechanism for the Emergence of Modern Culture and Behavior. *Biological Theory*, 13(4), 213–227.

d'Errico, F., Doyon, L., Colagé, I., et al. (2018). From Number Sense to Number Symbols: An Archaeological Perspective. *Philosophical Transactions of the Royal Society B: Biological Sciences*, 373(1740). doi:10.1098/RSTB.2016.0518

Deuker, L., Olligs, J., Fell, J., et al. (2013). Memory Consolidation by Replay of Stimulus-Specific Neural Activity. *The Journal of Neuroscience: The Official Journal of the Society for Neuroscience*, 33(49), 19373–19383.

Dickinson, A. (1985). Actions and Habits: The Development of Behavioural Autonomy. *Philosophical Transactions of the Royal Society of London. B, Biological Sciences*, 308(1135), 67–78.

Dickinson, A., & Balleine, B. (1994). Motivational Control of Goal-Directed Action. *Animal Learning & Behavior*, 22(1), 1–18.

Kendall, L. (n.d.). How to Deal with Hate on Social Media: Don't Feed the Trolls. https://blog.horizonsnhs.com/post/102fqwo/how-to-deal-with-hate-on-social-media-dont-feed-the-trolls

Doody, M., Van Swieten, M. M. H., & Manohar, S. G. (2022). Model-Based Learning Retrospectively Updates Model-Free Values. *Scientific Reports*, 12(1). doi:10.1038/S41598-022-05567-3

Dudai, Y., Karni, A., & Born, J. (2015). The Consolidation and Transformation of Memory. *Neuron*, 88(1), 20–32.

Dunbar, R. I. M. (1998). The Social Brain Hypothesis. *Evolutionary Anthropology: Issues, News, and Reviews*, 6(5), 178–190.

Earman, John. (1986). *A Primer on Determinism*, Dordrecht: Springer. Retrieved from https://link.springer.com/book/9789027722409

Edelmann, N. (2013). Reviewing the Definitions of "Lurkers" and Some Implications for Online Research. *Cyberpsychology, Behavior and Social Networking*, 16(9), 645–649.

Edney, S., Ryan, J. C., Olds, T., et al. (2019). User Engagement and Attrition in an App-Based Physical Activity Intervention: Secondary Analysis of a Randomized Controlled Trial. *J Med Internet Res 2019*, 21(11), e14645.

Ekman, P., & Davidson, R. (1994). *The Nature of Emotion: Fundamental Questions*, Oxford University Press.

Enfield, N. J., & Levinson, S. C. (2020). Introduction: Human Sociality As a New Interdisciplinary Field. In N. Enfield & S. Levinson (Eds.), *Roots of Human Sociality: Culture, Cognition and Interaction*, London: Routledge, 1–36.

Engström, P., & Forsell, E. (2018). Demand Effects of Consumers' Stated and Revealed Preferences. *Journal of Economic Behavior & Organization*, 150, 43–61.

Enriquez, J., & Gullans, S. (2015). *Evolving Ourselves: How Unnatural Selection and Nonrandom Mutation Are Changing Life on Earth*. London: OneWorld Publications.

Ersche, K. D., Meng, C., Ziauddeen, H., et al. (2020). Brain Networks Underlying Vulnerability and Resilience to Drug Addiction. *Proceedings of the National Academy of Sciences of the United States of America*, 117(26), 15253–15261.

Eshel, N. (2016). Trial and Error. *Science*, 354(6316), 1108–1109.

Estes, W. K. (1948). Discriminative Conditioning; Effects of a Pavlovian Conditioned Stimulus upon a Subsequently Established Operant Response. *Journal of Experimental Psychology*, 38(2), 173–177.

Ettinger, R., & Staddon, J. (1982). Behavior Competition, Component Duration and Multiple-Schedule Contrast. *Behaviour Analysis Letters*, 2(1), 31–38.

Feiner, L. (2023). Meta Sued by 42 Attorneys General Alleging Facebook, Instagram Features Are Addictive and Target Kids. Retrieved from www.cnbc.com/2023/10/24/bipartisan-group-of-ags-sue-meta-for-addictive-features.html

Ferster, C. B., & Skinner, B. F. (1957). *Schedules of Reinforcement*. East Norwalk: Appleton-Century-Crofts. doi:10.1037/10627-000

Finan, J. L., & Taylor, L. F. (1940). Quantitative Studies in Motivation. I. Strength of Conditioning in Rats under Varying Degrees of Hunger. *Journal of Comparative Psychology*, 29(1), 119–134.

Fineberg, N. A., Menchón, J. M., Hall, N., et al. (2022). Advances in Problematic Usage of the Internet Research: A Narrative Review by Experts from the European Network for Problematic Usage of the Internet. *Comprehensive Psychiatry*, 118. doi:10.1016/J.COMPPSYCH.2022.152346

Fisher, H. E., Aron, A., & Brown, L. L. (2006). Romantic Love: A Mammalian Brain System for Mate Choice. *Philosophical Transactions of the Royal Society of London. Series B, Biological Sciences*, 361(1476), 2173–2186.

Fiske, S. T. (2010). Interpersonal Stratification: Status, Power, and Subordination. In *Handbook of Social Psychology*, Hoboken, NJ, USA: John Wiley & Sons, Inc. doi:10.1002/9780470561119.socpsy002026

FitzGibbon, L., Lau, J. K. L., & Murayama, K. (2020). The Seductive Lure of Curiosity: Information As a Motivationally Salient Reward. *Current Opinion in Behavioral Sciences*, 35, 21–27.

Flavell, S. W., Gogolla, N., Lovett-Barron, M., & Zelikowsky, M. (2022). The Emergence and Influence of Internal States. *Neuron*, 110(16), 2545–2570.

Flayelle, M., Brevers, D., King, D. L., Maurage, P., Perales, J. C., & Billieux, J. (2023). A Taxonomy of Technology Design Features That Promote Potentially Addictive Online Behaviours. *Nature Reviews Psychology*, 2(3), 136–150.

Floccia, C., Christophe, A., & Bertoncini, J. (1997). High-Amplitude Sucking and Newborns: The Quest for Underlying Mechanisms. *Journal of Experimental Child Psychology*, 64(2), 175–198.

Flor, E. C. Á. (2019). La influencia del empiriocriticismo de E. Mach en la fundación de la psicología. *Discusiones Filosóficas*, 20(35), 91–110.

Fogg Behavior Model | Behavior Design Lab. (n.d.). Retrieved June 15, 2023, from https://behaviordesign.stanford.edu/resources/fogg-behavior-model

Forbes. (2023). How Ethical Business Tactics Can Improve Profitability. Retrieved June 27, 2023, from www.forbes.com/sites/forbesfinancecouncil/2 023/02/14/how-ethical-business-tactics-can-improve-profitability/? sh=68ce7480iddf

Foster, P. S., Hubbard, T., Campbell, R. W., et al. (2017a). Spreading Activation in Emotional Memory Networks and the Cumulative Effects of Somatic Markers. *Brain Informatics*, 4(2), 85–93.

Foster, P. S., Wakefield, C., Pryjmak, S., et al. (2017b). Spreading Activation in Nonverbal Memory Networks. *Brain Informatics*, 4(3), 187–199.

Foulkes, L., & Andrews, J. L. (2023). Are Mental Health Awareness Efforts Contributing to the Rise in Reported Mental Health Problems? A Call to Test the Prevalence Inflation Hypothesis. *New Ideas in Psychology*, 69, 101010.

Fragale, A. R., Overbeck, J. R., & Neale, M. A. (2011). Resources versus Respect: Social Judgments Based on Targets' Power and Status Positions. *Journal of Experimental Social Psychology*, 47(4), 767–775.

Frank, R. A., Hastings, L., & Stutz, R. M. (1984). Self-Deprivation: A Test of the Reward Hypothesis. *Physiology & Behavior*, 32(1), 139–141.

Frank, R. A., Pritchard, W. S., & Stutz, R. M. (1981). Food versus Intracranial Self-Stimulation: Failure of Limited-Access Self-Depriving Rats to Self-Deprive in a Continuous Access Paradigm. *Behavioral and Neural Biology*, 33(4), 503–508.

Fraser, K. M., Pribut, H. J., Janak, P. H., & Keiflin, R. (2023). From Prediction to Action: Dissociable Roles of Ventral Tegmental Area and Substantia Nigra Dopamine Neurons in Instrumental Reinforcement. *The Journal of Neuroscience: The Official Journal of the Society for Neuroscience*, 43(21), 3895–3908.

Friston, K. (2009). The Free-Energy Principle: A Rough Guide to the Brain? *Trends in Cognitive Sciences*, 13(7), 293–301.

Friston, K. (2010). The Free-Energy Principle: A Unified Brain Theory? *Nature Reviews Neuroscience*, 11(2), 127–138.

Fürnkranz, J., Chan, P. K., Craw, S., et al. (2011). Model-Based Reinforcement Learning. In *Encyclopedia of Machine Learning*, Boston, MA: Springer US, pp. 690–693.

Gamzu, E., & Schwartz, B. (1973). The Maintenance of Key Pecking by Stimulus-Contingent and Response-Independent Food Presentation. *Journal of the Experimental Analysis of Behavior*, 19(1), 65–72.

Garcia, B., Lebreton, M., Bourgeois-Gironde, S., & Palminteri, S. (2023). Experiential Values Are Underweighted in Decisions Involving Symbolic Options. *Nature Human Behaviour*, 7(4). doi:10.1038/S41562-022-01496-3

Garcí-Hoz, V. (2003). Signalization and Stimulus-Substitution in Pavlov's Theory of Conditioning. *The Spanish Journal of Psychology*, 6(2), 168–176.

Gershman, S. J., & Daw, N. D. (2017). Reinforcement Learning and Episodic Memory in Humans and Animals: An Integrative Framework. *Annual Review of Psychology*, 68, 101–128.

Gill, M., Sridhar, S., & Grewal, R. (2017). Return on Engagement Initiatives: A Study of a Business-to-Business Mobile App. *Journal of Marketing*, 81(4), 45–66.

Girskis, K. M., Stergachis, A. B., DeGennaro, E. M., et al. (2021). Rewiring of Human Neurodevelopmental Gene Regulatory Programs by Human Accelerated Regions. *Neuron*, 109(20), 3239–3251.e7.

Gläscher, J., Daw, N., Dayan, P., & O'Doherty, J. P. (2010). States versus Rewards: Dissociable Neural Prediction Error Signals Underlying Model-Based and Model-Free Reinforcement Learning. *Neuron*, 66(4), 585–595.

Gogoll, J., Zuber, N., Kacianka, S., Greger, T., Pretschner, A., & Nida-Rümelin, J. (2021). Ethics in the Software Development Process: From Codes of Conduct to Ethical Deliberation. *Philosophy & Technology*, 34(4), 1085–1108.

González-Forero, M., & Gardner, A. (2018). Inference of Ecological and Social Drivers of Human Brain-Size Evolution. *Nature*, 557(7706), 554–557.

Google AI. (n.d.). Google AI Principles. Retrieved June 27, 2023, from https://ai.google/responsibility/principles/

Gottlieb, J., & Oudeyer, P. Y. (2018). Towards a Neuroscience of Active Sampling and Curiosity. *Nature Reviews. Neuroscience*, 19(12), 758–770.

Gould, R. V., & Bearman, P. (2002). The Origins of Status Hierarchies: A Formal Theory and Empirical Test. *American Journal of Sociology*, 107(5), 1143–1178.

Grant, J. E., Potenza, M. N., Weinstein, A., & Gorelick, D. A. (2010). Introduction to Behavioral Addictions. *The American Journal of Drug and Alcohol Abuse*, 36(5), 233–241.

Greaves, S., & Fifer, S. (2010). Development of a Kilometer-Based Rewards System to Encourage Safer Driving Practices. *Transportation Research Record*, 2182, 88–96.

Greenspoon, J. (1955). The Reinforcing Effect of Two Spoken Sounds on the Frequency of Two Responses. *Am J Psychol*, 68(3), 409–416.

Gruber, C. W. (2014). Physiology of Invertebrate Oxytocin and Vasopressin Neuropeptides. *Experimental Physiology*, 99(1), 55–61.

Gruber, M. J., & Ranganath, C. (2019). How Curiosity Enhances Hippocampus-Dependent Memory: The Prediction, Appraisal, Curiosity, and Exploration (PACE) Framework. *Trends in Cognitive Sciences*, 23(12), 1014–1025.

Gruber, M. J., Ritchey, M., Wang, S. F., Doss, M. K., & Ranganath, C. (2016). Post-learning Hippocampal Dynamics Promote Preferential Retention of Rewarding Events. *Neuron*, 89(5), 1110.

Gründemann, J., Bitterman, Y., Lu, T., et al. (2019). Amygdala Ensembles Encode Behavioral States. *Science*, 364(6437). doi:10.1126/SCIENCE.AAV8736

Guilbeault, D., & Centola, D. (2020). Networked Collective Intelligence Improves Dissemination of Scientific Information Regarding Smoking Risks. *PLoS ONE*, 15(2). doi:10.1371/JOURNAL.PONE.0227813

Gulotta, K. B., & Byrne, T. (2015). A Progressive-Duration Schedule of Reinforcement. *Behavioural Processes*, 121, 93–97.

Gupta, S. (2016). Brain Food: Clever Eating. *Nature*, 531(7592), S12–13.

Halevy, N., Chou, E. Y., & Galinsky, A. D. (2011). A Functional Model of Hierarchy. *Organizational Psychology Review*, 1(1), 32–52.

Harlow, H. F. (1939). William James and Instinct Theory. In *William James: Unfinished Business*, Washington: American Psychological Association, pp. 21–30.

Hasan, H., & Kazlauskas, A. (2014). Activity Theory: Who Is Doing What, Why and How. In H. Hasan (Ed.), *Being Practical with Theory: A Window into Business Research* (pp. 9–14). Wollongong, Australia: THEORI. http://eureka connection.files.wordpress.com/2014/02/p-09-14-activity-theory-theori-eboo k-2014.pdf

Hayes, R. A., Carr, C. T., & Wohn, D. Y. (2016). One Click, Many Meanings: Interpreting Paralinguistic Digital Affordances in Social Media. *Journal of Broadcasting & Electronic Media*, 60(1), 171–187.

Hecht, E. E., Pargeter, J., Khreisheh, N., & Stout, D. (2023). Neuroplasticity Enables Bio-cultural Feedback in Paleolithic Stone-Tool Making. *Scientific Reports*, 13(1), 1–15.

Hellberg, S. N., Russell, T. I., & Robinson, M. J. F. (2019). Cued for Risk: Evidence for an Incentive Sensitization Framework to Explain the Interplay between Stress and Anxiety, Substance Abuse, and Reward Uncertainty in Disordered Gambling Behavior. *Cognitive, Affective & Behavioral Neuroscience*, 19(3), 737–758.

Henrich, J., Heine, S. J., & Norenzayan, A. (2010). The Weirdest People in the World? *The Behavioral and Brain Sciences*, 33(2–3), 61–83.

Henrich, J., & McElreath, R. (2003). The Evolution of Cultural Evolution. *Evolutionary Anthropology: Issues, News, and Reviews*, 12(3), 123–135.

Hentschel, T., Heilman, M. E., & Peus, C. V. (2019). The Multiple Dimensions of Gender Stereotypes: A Current Look at Men's and Women's Characterizations of Others and Themselves. *Frontiers in Psychology*, 10(JAN), 376558.

Herrmann, E., Keupp, S., Hare, B., Vaish, A., & Tomasello, M. (2013). Direct and Indirect Reputation Formation in Nonhuman Great Apes (Pan paniscus, Pan troglodytes, Gorilla gorilla, Pongo pygmaeus) and Human Children (Homo sapiens). *Journal of Comparative Psychology*, 127(1), 63–75.

Herrnstein, R. J. (1961). Relative and Absolute Strength of Response As a Function of Frequency of Reinforcement. *Journal of the Experimental Analysis of Behavior*, 4(3), 267–272.

Herrnstein, R. J. (1970). On the Law of Effect. *Journal of the Experimental Analysis of Behavior*, 13(2), 243–266.

Herrod, J. L., Snyder, S. K., Hart, J. B., Frantz, S. J., & Ayres, K. M. (2022). Applications of the Premack Principle: A Review of the Literature. *Behavior Modification*, 47(1), 219–246.

Heyes, C. (2014). *Cognitive Gadgets: The Cultural Evolution of Thinking. Antimicrobial Agents and Chemotherapy*, Vol. 58, Cambridge, Massachusetts: Harvard University Press. Retrieved from www.hup.harvard.edu/catalog.php? isbn=9780674980150

HFES. (n.d.). Code of Ethics. Retrieved June 27, 2023, from www.hfes.org/about-hfes/code-of-ethics

Hilbe, C., Chatterjee, K., & Nowak, M. A. (2018). Partners and Rivals in Direct Reciprocity. *Nature Human Behaviour*, 2(7), 469–477.

Hilgard, E. R. (1948). Thorndike's Connectionism. In *Theories of Learning*, East Norwalk: Appleton-Century-Crofts, pp. 19–51.

Hinde, R. (1953). Appetitive Behaviour, Consummatory Act, and the Hierarchical Organisation of Behaviour: With Special Reference to the Great Tit. *Behaviour*, 5(3), 189–224.

Hinson, J. M., & Staddon, J. E. R. (1978). Behavioral Competition: A Mechanism for Schedule Interactions. *Science*, 202(4366), 432–434.

Holland, P. C., & Rescorla, R. A. (1975). The Effect of Two Ways of Devaluing the Unconditioned Stimulus after First- and Second-Order Appetitive Conditioning. *Journal of Experimental Psychology. Animal Behavior Processes*, 1 (4), 355–363.

Hormes, J. M., Kearns, B., & Timko, C. A. (2014). Craving Facebook? Behavioral Addiction to Online Social Networking and Its Association with Emotion Regulation Deficits. *Addiction*, 109(12), 2079–2088.

Horváth, K., Nemeth, D., & Janacsek, K. (2022). Inhibitory Control Hinders Habit Change. *Scientific Reports*, 12(1). doi:10.1038/S41598-022-11971-6

Hosokawa, T., Kennerley, S. W., Sloan, J., & Wallis, J. D. (2013). Single-Neuron Mechanisms Underlying Cost-Benefit Analysis in Frontal Cortex. *Journal of Neuroscience*, 33(44), 17385–17397.

Huang, L., & Awh, E. (2018). Chunking in Working Memory via Content-Free Labels. *Scientific Reports*, 8(1). doi:10.1038/S41598-017-18157-5

Hughes, B. L., & Beer, J. S. (2012). Orbitofrontal Cortex and Anterior Cingulate Cortex Are Modulated by Motivated Social Cognition. *Cerebral Cortex*, 22(6), 1372–1381.

Hull, C. (1938). *Original Ideas on Things in General*, Yale University Library: Papers, Manuscripts and Archives.

Hull, C. (1943). *Principles of Behavior: An Introduction to Behavior Theory*. Appleton-Century. Retrieved from https://psycnet.apa.org/record/1944-00022-000

Hunt, L. T., Kolling, N., Soltani, A., Woolrich, M. W., Rushworth, M. F. S., & Behrens, T. E. J. (2012). Mechanisms Underlying Cortical Activity During Value-Guided Choice. *Nature Neuroscience*, 15(3), 470–476.

ILO. (n.d.). Principles and Guidelines for Human Factors / Ergonomics (HFE) Design and Management of Work Systems. Retrieved June 27, 2023, from www.ilo.org/global/topics/safety-and-health-at-work/news/WCMS_826603/lang–en/index.htm

Inman, J. J., & Nikolova, H. (2017). Shopper-Facing Retail Technology: A Retailer Adoption Decision Framework Incorporating Shopper Attitudes and Privacy Concerns. *Journal of Retailing*, 93(1), 7–28.

Inukollu, V. N., Keshamoni, D. D., Kang, T., & Inukollu, M. (2014). Factors Influencing Quality of Mobile Apps: Role of Mobile App Development Life Cycle. *Undefined*, 5(5), 15–34.

Izard, V., Dehaene-Lambertz, G., & Dehaene, S. (2008). Distinct Cerebral Pathways for Object Identity and Number in Human Infants. *PLoS Biology*, 6(2), 0275–0285.

Izuma, K., Saito, D. N., & Sadato, N. (2008). Processing of Social and Monetary Rewards in the Human Striatum. *Neuron*, 58(2), 284–294.

Jaffe, M. (2010). *The Primal Instinct*, Rowman & Littlefield Publishers.

Janofsky, M. (1993). Domino's Ends Fast-Pizza Pledge After Big Award to Crash Victim. *New York Times*. www.nytimes.com/1993/12/22/business/domino-s-en ds-fast-pizza-pledge-after-big-award-to-crash-victim.html#:~:text=But%20stu ng%20by%20a%20jury,longer%20promise%20such%20speedy%20delivery. &text=The%20company%20said%20it%20intended%20to%20appeal%20th e%20verdict

John, L. K., Loewenstein, G., Troxel, A. B., Norton, L., Fassbender, J. E., & Volpp, K. G. (2011). Financial Incentives for Extended Weight Loss: A Randomized, Controlled Trial. *Journal of General Internal Medicine*, 26(6), 621–626.

Kahneman, D. (2011). *Thinking, Fast and Slow*, New York: Farrar, Straus and Giroux.

Kahneman, D., & Tversky, A. (2018). Prospect Theory: An Analysis of Decision under Risk. *Experiments in Environmental Economics*, 1, 143–172.

Kensinger, E. A. (2009). Remembering the Details: Effects of Emotion. *Emotion Review: Journal of the International Society for Research on Emotion*, 1(2), 99–113.

Kensinger, E. A., Garoff-Eaton, R. J., & Schacter, D. L. (2007a). Effects of Emotion on Memory Specificity in Young and Older Adults. *The Journals of Gerontology. Series B, Psychological Sciences and Social Sciences*, 62(4), P208–15.

Kensinger, E. A., Garoff-Eaton, R. J., & Schacter, D. L. (2007b). How Negative Emotion Enhances the Visual Specificity of a Memory. *Journal of Cognitive Neuroscience*, 19(11), 1872–1887.

Keramati, M., & Gutkin, B. (2014). Homeostatic Reinforcement Learning for Integrating Reward Collection and Physiological Stability. *ELife*, 3. doi:10.7554/ ELIFE.04811

King, D., Delfabbro, P., & Griffiths, M. (2010). Video Game Structural Characteristics: A New Psychological Taxonomy. *International Journal of Mental Health and Addiction*, 8(1), 90–106.

Knudsen, E. B., & Wallis, J. D. (2021). Hippocampal Neurons Construct a Map of an Abstract Value Space. *Cell*, 184(18), 4640–4650.e10.

Kosinski, M., Stillwell, D., & Graepel, T. (2013). Private Traits and Attributes Are Predictable from Digital Records of Human Behavior. *Proceedings of the National Academy of Sciences of the United States of America*, 110(15), 5802–5805.

Koski, J. E., Xie, H., & Olson, I. R. (2015). Understanding Social Hierarchies: The Neural and Psychological Foundations of Status Perception. *Social Neuroscience*, 10(5), 527–550.

Krasner, L. (1958). Studies on the Conditioning of Verbal Behavior. *Psychological Bulletin*, 55(3), 148–170.

Kruger, J., & Evans, M. (2009). The Paradox of Alypius and the Pursuit of Unwanted Information. *Journal of Experimental Social Psychology*, 45(6), 1173–1179.

LaFontana, K. M., & Cillessen, A. H. N. (2010). Developmental Changes in the Priority of Perceived Status in Childhood and Adolescence. *Social Development*, 19(1), 130–147.

Laland, K. N., Odling-Smee, J., & Feldman, M. W. (2000). Niche Construction, Biological Evolution, and Cultural Change. *The Behavioral and Brain Sciences*, 23(1), 131–175.

Lally, P., Van Jaarsveld, C. H. M., Potts, H. W. W., & Wardle, J. (2010). How Are Habits Formed: Modelling Habit Formation in the Real World. *European Journal of Social Psychology*, 40(6), 998–1009.

Laraway, S., Snycerski, S., Olson, R., Becker, B., & Poling, A. (2014). The Motivating Operations Concept: Current Status and Critical Response. *The Psychological Record*, 64(3), 601–623.

Larche, C. J., Musielak, N., & Dixon, M. J. (2017). The Candy Crush Sweet Tooth: How "Near-Misses" in Candy Crush Increase Frustration, and the Urge to Continue Gameplay. *Journal of Gambling Studies*, 33(2), 599–615.

Lattal, K. A. (2010). Delayed Reinforcement of Operant Behavior. *Journal of the Experimental Analysis of Behavior*, 93(1), 129–139.

Lattal, K. A., & Gleeson, S. (1990). Response Acquisition With Delayed Reinforcement. *Journal of Experimental Psychology: Animal Behavior Processes*, 16(1), 27–39.

Lauer, D. (2021). Facebook's Ethical Failures Are Not Accidental; They Are Part of the Business Model. *AI and Ethics*, 1(4), 395–403.

Laursen, B., Altman, R. L., Bukowski, W. M., & Wei, L. (2020). Being Fun: An Overlooked Indicator of Childhood Social Status. *Journal of Personality*, 88(5), 993–1006.

Lavidge, R. J., & Steiner, G. A. (1961). A Model for Predictive Measurements of Advertising Effectiveness. *Undefined*, 25(6), 59–62.

Lawson, E. A., Marengi, D. A., Desanti, R. L., Holmes, T. M., Schoenfeld, D. A., & Tolley, C. J. (2015). Oxytocin Reduces Caloric Intake in Men. *Obesity*, 23(5), 950–956.

LeDoux, J. E., & Brown, R. (2017). A Higher-Order Theory of Emotional Consciousness. *Proceedings of the National Academy of Sciences*, 114(10). doi:10.1073/pnas.1619316114

Lee, J. C. (2021). Second-Order Conditioning in Humans. *Frontiers in Behavioral Neuroscience*, 15. doi:10.3389/FNBEH.2021.672628

Lee, Y., Jeon, Y. J., Kang, S., Shin, J. Il, Jung, Y. C., & Jung, S. J. (2022). Social Media Use and Mental Health During the COVID-19 Pandemic in Young Adults: A Meta-Analysis of 14 Cross-Sectional Studies. *BMC Public Health*, 22(1), 1–8.

León-Domínguez, U. (2022). *Conducta Virtual: Principios Básicos para el Diseño de Conductas Virtuales en el Metaverso*. Retrieved from www.amazon.es/Conducta-Virtual-Principios-Conductas-Virtuales/dp/B0B92HCPHH/ref=sr_1_1?__m

k_es_ES=%C3%85M%C3%85%C5%BD%C3%95%C3%91&crid=1C9DIV
No8X3MI&keywords=Conducta+Virtual&qid=1662555607&sprefix=conduct
a+virtual%2Caps%2C187&sr=8–1

León-Domínguez, U. (2024). Potential Cognitive Risks of Generative Transformer-Based AI Chatbots on Higher Order Executive Functions. *Neuropsychology*, 38(4), 293–308. doi:10.1037/neu0000948

León-Domínguez, U., Martín-Rodríguez, J. F., & León-Carrión, J. (2015). Executive n-back tasks for the neuropsychological assessment of working memory. *Behavioural Brain Research*, 292, 167–173.

Leont'ev, A. (1978). *Activity, Consciousness, and Personality*. Englewood Cliffs, NJ: Prentice-Hall. Retrieved from www.scirp.org/(S(i43dyn45teexjx455qlt3d2q))/r eference/ReferencesPapers.aspx?ReferenceID=1831210

Levy, D. J., & Glimcher, P. W. (2012). The Root of All Value: A Neural Common Currency for Choice. *Current Opinion in Neurobiology*, 22(6), 1027–1038.

Lewin, K. (1926). *Vorsatz und Beduerfniss*, Berlin: Springer.

Liddell, H. (1949). The Role of Vigilance in the Development of Animal Neurosis. *Proceedings of the Annual Meeting of the American Psychopathological Association*, 3, 183–196.

Lindström, B., Bellander, M., Schultner, D. T., Chang, A., Tobler, P. N., & Amodio, D. M. (2021). A computational reward learning account of social media engagement. *Nature Communications*, 12(1). doi:10.1038/S41467-020-19607-X

Liu, H., Lobschat, L., Verhoef, P. C., & Zhao, H. (2022). App Adoption: The Effect on Purchasing of Customers who have Used a Mobile Website Previously. *Journal of Interactive Marketing*, 47, 16–34.

Liu, J., Zhang, H., Yu, T., et al. (2021). Transformative Neural Representations Support Long-Term Episodic Memory. *Science Advances*, 7(41). doi:10.1126/SCIADV.ABG9715

Lovett-Barron, M., Andalman, A. S., Allen, W. E., et al. (2017). Ancestral Circuits for the Coordinated Modulation of Brain State. *Cell*, 171(6), 1411–1423.e17.

Lovibond, P. (1983). Facilitation of Instrumental Behavior by a Pavlovian Appetitive Conditioned Stimulus. *J Exp Psychol Anim Behav Process*, 9(3), 225–247.

Mackintosh, N. J. (1975). A Theory of Attention: Variations in the Associability of Stimuli with Reinforcement. *Psychological Review*, 82(4), 276–298.

Magee, J. C., & Galinsky, A. D. (2008). Social Hierarchy: The Self-Reinforcing Nature of Power and Status. *The Academy of Management Annals*, 2(1), 351–398.

Mar, A. C., & Robbins, T. W. (2007). Delay Discounting and Impulsive Choice in the Rat. *Current Protocols in Neuroscience*, Chapter 8: Unit 8.22. doi:10.1002/0471142301.NS0822S39

Márquez-Blanc, M. T. (2012). Causalidad, Identidad y Determinismo. *Revista de Filosofía*, 111–126.

Marr, D. (1982). *Vision: A Computational Investigation into the Human Representation and Processing of Visual Information*, San Francisco, CA: W.H. Freeman.

Martel, M., Cardinali, L., Roy, A. C., & Farnè, A. (2016). Tool-Use: An Open Window into Body Representation and Its Plasticity. *Cognitive Neuropsychology*, 33(1–2), 82–101.

Mason, A., Farrell, S., Howard-Jones, P., & Ludwig, C. J. H. (2017). The Role of Reward and Reward Uncertainty in Episodic Memory. *Journal of Memory and Language*, 96, 62–77.

Mather, M., & Nesmith, K. (2008). Arousal-Enhanced Location Memory for Pictures. *Journal of Memory and Language*, 58(2), 449–464.

Mazur, A. (1985). A Biosocial Model of Status in Face-to-Face Primate Groups. *Social Forces*, 64(2), 377.

Mazur, A. (2013). Biosocial Model of Status in Face-to-Face Primate Groups. *Procedia – Social and Behavioral Sciences*, 84, 53–56.

Mazur, J. E. (1997). Choice, Delay, Probability, and Conditioned Reinforcement. *Animal Learning & Behavior*, 25(2), 131–147.

McEwen, B. S., & Wingfield, J. C. (2010). What Is in a Name? Integrating Homeostasis, Allostasis and Stress. *Hormones and Behavior*, 57(2), 105–111.

McLean, G., Osei-Frimpong, K., Al-Nabhani, K., & Marriott, H. (2020). Examining Consumer Attitudes Towards Retailers' M-commerce Mobile Applications – An Initial Adoption vs. Continuous Use Perspective. *Journal of Business Research*, 106, 139–157.

McRae, K., Misra, S., Prasad, A. K., Pereira, S. C., & Gross, J. J. (2012). Bottom-up and Top-down Emotion Generation: Implications for Emotion Regulation. *Social Cognitive and Affective Neuroscience*, 7(3), 253–262.

McSweeney, F. K., & Weatherly, J. N. (1998). Habituation to the Reinforcer May Contribute to Multiple-Schedule Behavioral Contrast. *Journal of the Experimental Analysis of Behavior*, 69(2), 199–220.

Mead, G. (1934). *Mind, Self, and Society from the Standpoint of a Social Behaviorist*, Chicago: University of Chicago Press.

Mead, G. H. (1923). Scientific Method and the Moral Sciences. *International Journal of Ethics*, 33(3), 229–247.

Meta. (n.d.). The Code of Conduct. Retrieved June 27, 2023, from https://about .meta.com/code-of-conduct/

Michael, J., Palmer, D. C., & Sundberg, M. L. (2011). The Multiple Control of Verbal Behavior. *The Analysis of Verbal Behavior*, 27(1), 3–22.

Microsoft Legal. (n.d.). Legal Compliance and Ethics. Retrieved June 27, 2023, from www.microsoft.com/en-us/legal/compliance

Miller, E. K., & Cohen, J. D. (2001). An Integrative Theory of Prefrontal Cortex Function. *Annual Review of Neuroscience*, 24(1), 167–202.

Miller, G. A., Galanter, E., & Pribram, K. H. (1960). *Plans and the Structure of Behavior*, New York: Henry Holt and Co. doi:10.1037/10039-000

Miller, K., & Paredes, D. (1996). *The Nature of Mathematical Thinking*, New York and London: Routledge. Retrieved from https://books.google.com.mx/books? hl=es&lr=&id=SX8pPNs3PCkC&oi=fnd&pg=PA83&ots=XdOJRYaTok&si g=Lpgg3Mg4vKyPYYu6faHclcCdMz4&redir_esc=y#v=snippet&q=numerical %20cognition&f=false

Miller, L. E., Fabio, C., Ravenda, V., et al. (2019). Somatosensory Cortex Efficiently Processes Touch Located Beyond the Body. *Current Biology: CB*, 29(24), 4276–4283.e5.

Miller, L. E., Longo, M. R., & Saygin, A. P. (2014). Tool Morphology Constrains the Effects of Tool Use on Body Representations. *Journal of Experimental Psychology. Human Perception and Performance*, 40(6), 2143–2153.

Miller, L. E., Montroni, L., Koun, E., Salemme, R., Hayward, V., & Farnè, A. (2018). Sensing with Tools Extends Somatosensory processing beyond the body. *Nature*, 561(7722), 239–242.

Miller, N. (1951). Learnable Drives and Rewards. In *Handbook of Experimental Psychology*, New York: Wiley, pp. 435–472.

Miller, N. E. (1948). Studies of Fear As an Acquirable Drive Fear As Motivation and Fear-Reduction As Reinforcement in the Learning of New Responses. *Journal of Experimental Psychology*, 38(1), 89–101.

Minamimoto, T., La Camera, G., & Richmond, B. J. (2009). Measuring and Modeling the Interaction among Reward Size, Delay to Reward, and Satiation Level on Motivation in Monkeys. *Journal of Neurophysiology*, 101(1), 437–447.

Moll, H., & Tomasello, M. (2007). Cooperation and Human Cognition: The Vygotskian Intelligence Hypothesis. *Philosophical Transactions of the Royal Society of London. Series B, Biological Sciences*, 362(1480), 639–648.

Morf, M. E., & Weber, W. G. (2000). I/O Psychology and the Bridging of A. N. Leont'ev's Activity Theory. *Canadian Psychology/Psychologie Canadienne*, 41(2), 81–93. https://doi.org/10.1037/h0088234

Morrens, J., Aydin, Ç., Janse van Rensburg, A., Esquivelzeta Rabell, J., & Haesler, S. (2020). Cue-Evoked Dopamine Promotes Conditioned Responding during Learning. *Neuron*, 106(1), 142–153.e7.

Morris, L. S., Grehl, M. M., Rutter, S. B., Mehta, M., & Westwater, M. L. (2022). On What Motivates Us: A Detailed Review of Intrinsic *v.* Extrinsic Motivation. *Psychological Medicine*, 52(10), 1801–1816.

Müller, A., Brand, M., Claes, L., et al. (2019). Buying-Shopping Disorder – Is There Enough Evidence to Support Its Inclusion in ICD-11? *CNS Spectrums*, 24 (4), 374–379.

Müller, V. C. (2020). *Risks of Artificial Intelligence*, Boca Raton: CRC PRESS. Retrieved from www.routledge.com/Risks-of-Artificial-Intelligence/Muller/p/book/9780367575182

Muthukrishna, M., Doebeli, M., Chudek, M., & Henrich, J. (2018). The Cultural Brain Hypothesis: How Culture Drives Brain Expansion, Sociality, and Life History. *PLoS Computational Biology*, 14(11). doi:10.1371/JOURNAL.PCBI.1006504

Muthukrishna, M., & Henrich, J. (2016). Innovation in the Collective Brain. *Philosophical Transactions of the Royal Society of London. Series B, Biological Sciences*, 371(1690). doi:10.1098/RSTB.2015.0192

New, J., Cosmides, L., & Tooby, J. (2007). Category-Specific Attention for Animals Reflects Ancestral Priorities, Not Expertise. *Proceedings of the National Academy of Sciences of the United States of America*, 104(42), 16598–16603.

NiemanLab. (2018). What Happens When Facebook Goes Down? People Read the News. Retrieved June 26, 2023, from www.niemanlab.org/2018/10/what-happens-when-facebook-goes-down-people-read-the-news/

O'Doherty, J. P., Cockburn, J., & Pauli, W. M. (2017). Learning, Reward, and Decision Making. *Annual Review of Psychology*, 68, 73–100.

Olds, J., & Milner, P. (1954). Positive Reinforcement Produced by Electrical Stimulation of Septal Area and Other Regions of Rat Brain. *Journal of Comparative and Physiological Psychology*, 47(6), 419–427.

O'Lemmon, M. (2020). The Technological Singularity as the Emergence of a Collective Consciousness: An Anthropological Perspective. *Bulletin of Science, Technology & Society*, 40(1–2), 15–27.

Oosterwijk, S. (2017). Choosing the Negative: A Behavioral Demonstration of Morbid Curiosity. *PLOS ONE*, 12(7), e0178399.

Ormaxabal, K. M. (2006). The Law of Diminishing Marginal Utility in Alfred Marshall's Principles of Economics. *The European Journal of the History of Economic Thought*, 2(1), 91–126.

Ortega y Gasset, J. (n.d.). Thoughts on Technology. In *Philosophy and Technology – Readings in the Philosophical Problems of Technology*, New York: Free Press, pp. 290–313.

Ozer, D. J., & Benet-Martínez, V. (2006). Personality and the Prediction of Consequential Outcomes. *Annual Review of Psychology*, 57, 401–421.

Page, S., & Neuringer, A. (1985). Variability Is an Operant. *Journal of Experimental Psychology: Animal Behavior Processes*, 11(3), 429–452.

Pani, L. (2000). Is There an Evolutionary Mismatch between the Normal Physiology of the Human Dopaminergic System and Current Environmental Conditions in Industrialized Countries? *Molecular Psychiatry*, 5(5), 467–475.

Panova, T., & Carbonell, X. (2018). Is Smartphone Addiction Really an Addiction? *Journal of Behavioral Addictions*, 7(2), 252–259.

Paulas, R. (2017). Do Ethics Have a Place in Business Schools? https://psmag.com/economics/millennials-youre-our-only-hope

Pavlov, I. (1927). *Conditioned Reflexes: An Investigation of the Physiological Activity of the Cerebral Cortex*, Oxford University Press.

Pelletier, G., Aridan, N., Fellows, L. K., & Schonberg, T. (2021). A Preferential Role for Ventromedial Prefrontal Cortex in Assessing "the Value of the Whole" in Multiattribute Object Evaluation. *The Journal of Neuroscience: The Official Journal of the Society for Neuroscience*, 41(23), 5056–5068.

Petry, N. (2015). *Behavioral Addictions: DSM-5® and Beyond*, Oxford University Press. Retrieved from https://global.oup.com/academic/product/behavioral-addictions-dsm-5-and-beyond-9780199391547?cc=mx&lang=en&

Petta Gomes da Costa, D. L. (2019). Reviewing the Concept of Technological Singularities: How Can It Explain Human Evolution? *NanoEthics*, 13(2), 119–130.

Pfeiffer, B. E., & Foster, D. J. (2013). Hippocampal Place-Cell Sequences Depict Future Paths to Remembered Goals. *Nature*, 497(7447), 74–79.

Philiastides, M. G., Biele, G., & Heekeren, H. R. (2010). A Mechanistic Account of Value Computation in the Human Brain. *Proceedings of the National Academy of Sciences of the United States of America*, 107(20), 9430–9435.

Piazza, C. C., Bowman, L. G., Contrucci, S. A., Delia, M. D., Adelinis, J. D., & Goh, H.-L. (1999). An Evaluation of the Properties of Attention As Reinforcement for Destructive and Appropriate Behavior. *Journal of Applied Behavior Analysis*, 32(4), 437–449.

Piccinini, G. (2020). Nonnatural Mental Representation. In *What Are Mental Representations?*, Oxford University Press, pp. 254–286.

Piccinini, G. (2022). Situated Neural Representations: Solving the Problems of Content. *Frontiers in Neurorobotics*, 16, 73.

Pickens, C. L., & Holland, P. C. (2004). Conditioning and Cognition. *Neuroscience & Biobehavioral Reviews*, 28(7), 651–661.

Pinker, S. (2010). The Cognitive Niche: Coevolution of Intelligence, Sociality, and Language. *Proceedings of the National Academy of Sciences*, 107(supplement_2), 8993–8999.

Poldrack, R. A. (2021). *Hard to Break: Why Our Brains Make Habits Stick*. Princeton: Princeton University Press. doi:10.2307/J.CTV191KX44

Popper, K. (1965). Time's Arrow and Entropy. *Nature*, 207(4994), 233–234.

Powers, A. R., Mathys, C., & Corlett, P. R. (2017). Pavlovian Conditioning-Induced Hallucinations Result from Overweighting of Perceptual Priors. *Science*, 357(6351), 596–600.

Pozo, M., Milà-Guasch, M., Haddad-Tóvolli, R., et al. (2023). Negative Energy Balance Hinders Prosocial Helping Behavior. *Proceedings of the National Academy of Sciences*, 120(15). doi:10.1073/pnas.2218142120

Raghunathan, R., & Pham, M. (1999). All Negative Moods Are Not Equal: Motivational Influences of Anxiety and Sadness on Decision Making. *Organizational Behavior and Human Decision Processes*, 79(1), 56–77.

Ramsay, D. S., & Woods, S. C. (2014). Clarifying the Roles of Homeostasis and Allostasis in Physiological Regulation. *Psychological Review*, 121(2), 225–247.

Randjelović, P., Stojiljković, N., Radulović, N., Ilić, I., Stojanović, N., & Ilić, S. (2018). The Association of Smartphone Usage with Subjective Sleep Quality and Daytime Sleepiness Among Medical Students. *Biological Rhythm Research*, 50(6), 857–865.

Rangel, A., Camerer, C., & Montague, P. R. (2008). A Framework for Studying the Neurobiology of Value-Based Decision Making. *Nature Reviews. Neuroscience*, 9(7), 545–556.

Rao, R. P. N., & Ballard, D. H. (1999). Predictive Coding in the Visual Cortex: A Functional Interpretation of Some Extra-Classical Receptive-Field Effects. *Nature Neuroscience*, 2(1), 79–87.

Raza, S. H., Yousaf, M., Sohail, F., et al. (2021). Investigating Binge-Watching Adverse Mental Health Outcomes During Covid-19 Pandemic: Moderating Role of Screen Time for Web Series Using Online Streaming. *Psychol Res Behav Manag.*, 14, 1615–1629.

Reddy, L., Poncet, M., Self, M. W., et al. (2015). Learning of Anticipatory Responses in Single Neurons of the Human Medial Temporal Lobe. *Nature Communications*, 6(1), 8556.

Reed, P., Fowkes, T., & Khela, M. (2023). Reduction in Social Media Usage Produces Improvements in Physical Health and Wellbeing: An RCT. *Journal of Technology in Behavioral Science*, 8(2), 140–147.

Reeh, P. W., & Fischer, M. J. M. (2022). Nobel Somatosensations and Pain. *Pflugers Archiv: European Journal of Physiology*, 474(4), 405–420.

Reiss, S. (2000). *Who Am I: The 16 Basic Desires That Motivate Our Actions and Define Our Personality*, New York: Putnam.

Rescorla, R. A. (1973). Effect of US Habituation Following Conditioning. *Journal of Comparative and Physiological Psychology*, 82(1), 137–143.

Rescorla, R. A. (1974). Effect of Inflation of the Unconditioned Stimulus Value Following Conditioning. *Journal of Comparative and Physiological Psychology*, 86 (1), 101–106.

Rescorla, R. A., & Cunningham, C. L. (1978). Within-Compound Flavor Associations. *Journal of Experimental Psychology. Animal Behavior Processes*, 4 (3), 267–275.

Rescorla, R. A., & Lolordo, V. M. (1965). Inhibition of Avoidance Behavior. *Journal of Comparative and Physiological Psychology*, 59(3), 406–412.

Rescorla, R. A., & Solomon, R. L. (1967). Two-Process Learning Theory: Relationships between Pavlovian Conditioning and Instrumental Learning. *Psychological Review*, 74(3), 151–182.

Rescorla, R., & Wagner, A. (1972). A Theory of Pavlovian Conditioning: Variations in the Effectiveness of Reinforcement and Nonreinforcement. In *Classical Conditioning II: Current Research and Theory*, New York: Appleton-Century-Crofts, pp. 64–99.

Reynolds, G. S. (1961). Behavioral Contrast. *Journal of the Experimental Analysis of Behavior*, 4(1), 57–71.

Richerson, P. J., & Boyd, R. (1978). A Dual Inheritance Model of the Human Evolutionary Process I: Basic Postulates and a Simple Model. *Journal of Social and Biological Structures*, 1(2), 127–154.

Rilling, J. K. (2014). Comparative Primate Neurobiology and the Evolution of Brain Language Systems. *Current Opinion in Neurobiology*, 28, 10–14.

Roberts, W. A. (2014). Instrumental and Classical Conditioning. *The Wiley Blackwell Handbook of Operant and Classical Conditioning*, Malden, MA: Wiley Blackwell, 417–451.

Robertson, T. E., Sznycer, D., Delton, A. W., Tooby, J., & Cosmides, L. (2018). The True Trigger of Shame: Social Devaluation Is Sufficient, Wrongdoing Is Unnecessary. *Evolution and Human Behavior*, 39(5), 566–573.

Robinson, M. J. F., Clibanoff, C., Freeland, C. M., Knes, A. S., Cote, J. R., & Russell, T. I. (2019). Distinguishing between Predictive and Incentive Value of Uncertain Gambling-Like Cues in a Pavlovian Autoshaping Task. *Behavioural Brain Research*, 371. doi:10.1016/J.BBR.2019.111971

Robinson, T. E., & Berridge, K. C. (2008). Review. The Incentive Sensitization Theory of Addiction: Some Current Issues. *Philosophical Transactions of the Royal Society of London. Series B, Biological Sciences*, 363 (1507), 3137–3146.

Rodriguez Cabrero, J. A. M., Zhu, J. Q., & Ludvig, E. A. (2019). Costly Curiosity: People Pay a Price to Resolve an Uncertain Gamble Early. *Behavioural Processes*, 160, 20–25.

Rolls, E. T. (2004). The Functions of the Orbitofrontal Cortex. *Brain and Cognition*, 55(1), 11–29.

Rosati, A. G. (2017). Foraging Cognition: Reviving the Ecological Intelligence Hypothesis. *Trends in Cognitive Sciences*, 21(9), 691–702.

Rosenberg, K. P., & Feder, L. C. (2014). An Introduction to Behavioral Addictions. In K. P. Rosenberg & L. Curtiss Feder (Eds.), *Behavioral Addictions: Criteria, Evidence, and Treatment*, Oxford: Elsevier Academic Press, 1–17.

Rossi, R. R., & Stutz, R. M. (1978). The Self-Deprivation Phenomenon: Competition between Appetitive Rewards and Electrical Stimulation of the Brain. *Physiological Psychology*, 6(2), 204–208.

Rostami Kandroodi, M., Cook, J. L., Swart, J. C., et al. (2021). Effects of Methylphenidate on Reinforcement Learning Depend on Working Memory Capacity. *Psychopharmacology*, 238(12), 3569–3584.

Routtenberg, A., & Lindy, J. (1965). Effects of the Availability of Rewarding Septal and Hypothalamic Stimulation on Bar Pressing for Food Under Conditions of Deprivation. *Journal of Comparative and Physiological Psychology*, 60(2), 158–161.

Ryan, R. M., & Deci, E. L. (2000). Self-Determination Theory and the Facilitation of Intrinsic Motivation, Social Development, and Well-Being. *The American Psychologist*, 55(1), 68–78.

Safronova, V. (2017). The Rise and Fall of Yik Yak, the Anonymous Messaging App. Retrieved June 11, 2023, from www.nytimes.com/2017/05/27/style/yik-yak-bullying-mary-washington.html

Salthouse, T. A., & Pink, J. E. (2008). Why Is Working Memory Related to Fluid Intelligence? *Psychonomic Bulletin & Review*, 15(2), 364–371.

Sapolsky, R. M. (2005). The Influence of Social Hierarchy on Primate Health. *Science*, 308(5722), 648–652.

Savin-Williams, R. C. (1979). Dominance Hierarchies in Groups of Early Adolescents. *Child Development*, 50(4), 923.

Schiavone, G., Fallegger, F., Kang, X., et al. (2020). Soft, Implantable Bioelectronic Interfaces for Translational Research. *Advanced Materials*, 32 (17), 1906512.

Schlaudt, O. (2022). Exaptation in the Co-evolution of Technology and Mind: New Perspectives from Some Old Literature. *Philosophy and Technology*, 35(2), 1–26.

Schlichtinga, M. L., & Prestona, A. R. (2014). Memory Reactivation During Rest Supports Upcoming Learning of Related Content. *Proceedings of the National Academy of Sciences of the United States of America*, 111(44), 15845–15850.

Schwartz, M. (2023). Adaptive Ethics for Digital Transformation: A New Approach for Enterprise Leaders (eBook). IT Revolution Press.

Schwartz, M. S. (2017). Teaching Behavioral Ethics: Overcoming the Key Impediments to Ethical Behavior. *Journal of Management Education*, 41(4), 497–513.

Schwartz, S. H. (1992). Universals in the Content and Structure of Values: Theoretical Advances and Empirical Tests in 20 Countries. *Advances in Experimental Social Psychology*, 25, 1–65.

Schwarz, A. F., Huertas-Delgado, F. J., Cardon, G., & Desmet, A. (2020). Design Features Associated with User Engagement in Digital Games for Healthy Lifestyle Promotion in Youth: A Systematic Review of Qualitative and Quantitative Studies. *Games for Health Journal*, 9(3), 150–163.

Sennesh, E., Theriault, J., Brooks, D., van de Meent, J.-W., Barrett, L. F., & Quigley, K. S. (2022). Interoception As Modeling, Allostasis As Control. *Biological Psychology*, 167, 108242.

Sheffield, F. D., & Roby, T. B. (1950). Reward Value of a Non-Nutritive Sweet-Taste. *Journal of Comparative and Physiological Psychology*, 43(6), 471–481.

Sheffield, F. D., Roby, T. B., & Campbell, B. A. (1954). Drive Reduction versus Consummatory Behavior As Determinants of Reinforcement. *Journal of Comparative and Physiological Psychology*, 47(5), 349–354.

Sheffield, F. D., Wulff, J. J., & Backer, R. (1951). Reward Value of Copulation Without Sex Drive Reduction. *Journal of Comparative and Physiological Psychology*, 44(1), 3–8.

Sheth, J. N., Newman, B. I., & Gross, B. L. (1991). Why We Buy What We Buy: A Theory of Consumption Values. *Journal of Business Research*, 22(2), 159–170.

Shikano, Y., Ikegaya, Y., & Sasaki, T. (2021). Minute-Encoding Neurons in Hippocampal-Striatal Circuits. *Current Biology*, 31(7), 1438–1449.e6.

Shimp, C. P. (2013). Toward the Unification of Molecular and Molar Analyses. *The Behavior Analyst*, 36(2), 295–312.

Silverman, K., DeFulio, A., & Sigurdsson, S. O. (2012). Maintenance of Reinforcement to Address the Chronic Nature of Drug Addiction. *Preventive Medicine*, 55 (Suppl). doi:10.1016/J.YPMED.2012.03.013

Simon, H. (1957). *Models of Man: Social and Rational*, New York: Wiley.

Skinner, B. (1974). *About Behaviorism*, New York: Vintage Books.

Skinner, B. (1938). *The Behavior of Organisms: An Experimental Analysis*. PsycNET, Appleton-Century. Retrieved from https://psycnet.apa.org/record/1939-00056-000

Skinner, B. (1970). *Ciencia y conducta humana*, Barcelona: Fontanella. Retrieved from http://chamilo.cut.edu.mx:8080/chamilo/courses/FUNDAMENTOSEPISTEMOLOGICOSDELAPSI2/document/3._ciencia_conducta_humana_skinner.pdf

Plazas, E. A. (2006). B. F. Skinner: la búsqueda de orden en la conducta voluntaria. *Universitas Psychologica*, 5(2): 371–383.

Spier, L. (1971). A Suggested Behavioral Approach to Cost-Benefit Analysis. *Management Science*, 672–693.

Spreckelmeyer, K. N., Krach, S., Kohls, G., et al. (2009). Anticipation of Monetary and Social Reward Differently Activates Mesolimbic Brain Structures in Men and Women. *Social Cognitive and Affective Neuroscience*, 4 (2), 158–165.

Squire, L. R. (2004). Memory Systems of the Brain: A Brief History and Current Perspective. *Neurobiology of Learning and Memory*, 82(3), 171–177.

Staddon, J. (2003). *Adaptive Behavior and Learning*, London and New York: Cambridge University Press. Retrieved from https://moam.info/adaptive-beha vior-and-learning_5c177d67097c4784128b4659.html

Stark, R., Klucken, T., Potenza, M. N., Brand, M., & Strahler, J. (2018). A Current Understanding of the Behavioral Neuroscience of Compulsive Sexual Behavior Disorder and Problematic Pornography Use. *Current Behavioral Neuroscience Reports*, 5(4), 218–231.

Stein, L., Xue, B. G., & Belluzzi, J. D. (1994). In Vitro Reinforcement of Hippocampal Bursting: A Search for Skinner's Atoms of Behavior. *Journal of the Experimental Analysis of Behavior*, 61(2), 155–168.

Sterelny, K. (2020). Afterword: Tough Questions; Hard Problems; Incremental Progress. *Topics in Cognitive Science*, 12(2), 766–783.

Stice, E., Burger, K., & Yokum, S. (2013). Caloric Deprivation Increases Responsivity of Attention and Reward Brain Regions to Intake, Anticipated Intake, and Images of Palatable Foods. *NeuroImage*, 67, 322–330.

Stigler, G. J. (1950). The Development of Utility Theory. I. *Journal of Political Economy*, 58(4), 307–327.

Suddendorf, T., Bulley, A., & Miloyan, B. (2018). Prospection and Natural Selection. *Current Opinion in Behavioral Sciences*, 24, 26–31.

Sumner, R., Burrow, A. L., & Hill, P. L. (2014). Identity and Purpose as Predictors of Subjective Well-Being in Emerging Adulthood. http://Dx.Doi.Org/10.1177/ 2167696814532796, 3(1), 46–54.

Sutphin, G., Byrne, T., & Poling, A. (1998). Response Acquisition with Delayed Reinforcement: A Comparison of Two-Lever Procedures. *Journal of the Experimental Analysis of Behavior*, 69(1), 17–28.

Tangney, J. P., Baumeister, R. F., & Boone, A. L. (2004). High Self-Control Predicts Good Adjustment, Less Pathology, Better Grades, and Interpersonal Success. *Journal of Personality*, 72(2), 271–324.

Taylor, J. C. (1949). Behavior Oscillation and the Growth of Preference. *Psychological Review*, 56(2), 77–87.

TechCrunch. (2018). Stories Are About to Surpass Feed Sharing. Now What? | TechCrunch. Retrieved June 26, 2023, from https://techcrunch.com/2018/05/ 02/stories-are-about-to-surpass-feed-sharing-now-what/

Tennie, C., Call, J., & Tomasello, M. (2009). Ratcheting up the Ratchet: On the Evolution of Cumulative Culture. *Philosophical Transactions of the Royal Society of London. Series B, Biological Sciences*, 364(1528), 2405–2415.

Tesch, A. D., & Sanfey, A. G. (2008). Models and Methods in Delay Discounting. *Annals of the New York Academy of Sciences*, 1128, 90–94.

Trotzke, P., Starcke, K., Müller, A., et al. (2015). Pathological Buying Online as a Specific Form of Internet Addiction: A Model-Based Experimental Investigation. *PLoS One*, 10:e0140296.

Thaler, R., & Sunstein, C. (2009). *Nudge. Improving Decisions About Health, Wealth, And Happiness*, London: Penguin.

The New Statesman. (2021). The Great Facebook Outage of 2021: Why WhatsApp and Instagram Were Down for Six Hours – New Statesman. Retrieved June 26, 2023, from www.newstatesman.com/culture/social-media/2021/10/the-great-facebook-outage-of-2021-why-whatsapp-and-instagram-were-down-for-six-hours-yesterday

The Oxford Dictionary of Sports Science & Medicine. (2006). *The Oxford Dictionary of Sports Science & Medicine*. doi:10.1093/ACREF/9780198568068.001.0001

Thorndike, E. (1911). *Animal Intelligence*, New York: McMillan. Retrieved from https://openlibrary.org/books/OL7172495M/Animal_intelligence

Thorndike, E. L. (1905). Measurement of Twins. *The Journal of Philosophy, Psychology and Scientific Methods*, 2(20), 547–553.

Tik, M., Sladky, R., Luft, C. D. B., et al. (2018). Ultra-High-Field fMRI Insights on Insight: Neural Correlates of the Aha!-Moment. *Human Brain Mapping*, 39(8), 3241–3252.

Tinbergen, N. (1951). *The Study of Instinct*, Oxford: Clarendon Press/Oxford University Press.

Tipoe, E., Adams, A., & Crawford, I. (2022). Revealed Preference Analysis and Bounded Rationality. *Oxford Economic Papers*, 74(2), 313–332.

Tomasello, M., & Carpenter, M. (2005). The Emergence of Social Cognition in Three Young Chimpanzees. *Monographs of the Society for Research in Child Development*, 70(1), vii–132.

Tonegawa, S., Pignatelli, M., Roy, D. S., & Ryan, T. J. (2015). Memory Engram Storage and Retrieval. *Current Opinion in Neurobiology*, 35, 101–109.

Tonneau, F. (2012). Associationism. In *Encyclopedia of the Sciences of Learning*, Boston, MA: Springer US, pp. 326–329.

Tulving, E. (1983). *Elements of Episodic Memory*, Oxford University Press.

US Bureau of Labor Statistics. (2022). Table 1. Time spent in primary activities and percent of the civilian population engaging in each activity, averages per day by sex, 2022 annual averages – 2022 A01 Results. Retrieved June 26, 2023, from www.bls.gov/news.release/atus.t01.htm.

van der Burg, S., & Swierstra, T. (2013). Ethics on the Laboratory Floor. *Ethics on the Laboratory Floor*, 1–230.

van Heerde, H. J., Dinner, I. M., & Neslin, S. A. (2019). Engaging the Unengaged Customer: The Value of a Retailer Mobile App. *International Journal of Research in Marketing*, 36(3), 420–438.

van Noort, G., & van Reijmersdal, E. A. (2022). Branded Apps: Explaining Effects of Brands' Mobile Phone Applications on Brand Responses. *Journal of Interactive Marketing*, 45, 16–26.

van Schaik, C. P., & Burkart, J. M. (2011). Social Learning and Evolution: The Cultural Intelligence Hypothesis. *Philosophical Transactions of the Royal Society of London. Series B, Biological Sciences*, 366(1567), 1008–1016.

Vogel, E. H., & Wagner, A. R. (2018). SOP Model. *Encyclopedia of Animal Cognition and Behavior*, Princeton: Princeton University Press, 1–3.

von Neumann, J., Morgenstern, O., Kuhn, H. W., & Rubinstein, A. (1944). Formulation of The Economic Problem. *Theory of Games and Economic Behavior (60th Anniversary Commemorative Edition)*, 1–45.

Vygotsky, L. (1962). *Thought and Language*, Cambridge, Mass: MIT Press.

Vygotsky, L. (1978). *Mind in Society: The Development of Higher Psychological Processes*. Cambridge, Mass: Harvard University Press.

Wakker, P., & Tversky, A. (1993). An Axiomatization of Cumulative Prospect Theory. *Journal of Risk and Uncertainty*, 7(2), 147–175.

Waltz, T. J., & Follette, W. C. (2009). Molar Functional Relations and Clinical Behavior Analysis: Implications for Assessment and Treatment. *The Behavior Analyst*, 32(1), 51–68.

Wang, M., Han, Q., Gui, C., et al. (2019). Differences in the Risk Assessment of SoilHeavy Metals between Newly Built and Original Parks in Jiaozuo, Henan Province, China. *The Science of the Total Environment*, 676, 1–10.

Wang, S., Chen, X., Zhao, C., et al. (2023). An Organic Electrochemical Transistor for Multi-Modal Sensing, Memory and Processing. *Nature Electronics*, 6(4), 281–291.

Ware, M. (1999). *The History, Science and Art of Photographic Printing in Prussian Blue*, Science Museum and National Museum of Photography, Film & Television.

Waring, T. M., & Wood, Z. T. (2021). Long-Term Gene–Culture Coevolution and the Human Evolutionary Transition. *Proceedings of the Royal Society B*, 288 (1952). doi:10.1098/RSPB.2021.0538

Warneken, F., & Tomasello, M. (2006). Altruistic Helping in Human Infants and Young Chimpanzees. *Science*, 311(5765), 1301–1303.

Wegner, D. M., Schneider, D. J., Carter, S. R., & White, T. L. (1987). Paradoxical Effects of Thought Suppression. *Journal of Personality and Social Psychology*, 53 (1), 5–13.

Weiss, S. (2014). The Instrumentally Derived Incentive-Motivational Function. *International Journal of Comparative Psychology*, 27(4), 598–613.

Weiss, S. J. (1978). Discriminated Response and Incentive Processes in Operant Conditioning: A Two-Factor Model of Stimulus Control. *Journal of the Experimental Analysis of Behavior*, 30(3), 361–381.

Weissbourd, B., Momose, T., Nair, A., Kennedy, A., Hunt, B., & Anderson, D. J. (2021). A Genetically Tractable Jellyfish Model for Systems and Evolutionary Neuroscience. *Cell*, 184(24), 5854–5868.e20.

Welbers, K., Konijn, E. A., Burgers, C., de Vaate, A. B., Eden, A., & Brugman, B. C. (2019). Gamification As a Tool for Engaging Student Learning: A Field Experiment with a Gamified App. *E-learning and Digital Media*, 16(2), 92–109. https://Doi.Org/10.1177/2042753018818342.

Williams, E. H., Bilbao-Broch, L., Downing, P. E., & Cross, E. S. (2020). Examining the Value of Body Gestures in Social Reward Contexts. *NeuroImage*, 222. doi:10.1016/J.NEUROIMAGE.2020.117276

Williams, W. A. (2019). Innate Releasing Mechanism. *Encyclopedia of Animal Cognition and Behavior*, 1–6.

Wilmer, H. H., Sherman, L. E., & Chein, J. M. (2017). Smartphones and Cognition: A Review of Research Exploring the Links between Mobile Technology Habits and Cognitive Functioning. *Frontiers in Psychology*, 8 (APR). doi:10.3389/FPSYG.2017.00605

Windholz, G. (1983). Pavlov's Position toward American Behaviorism. *Journal of the History of the Behavioral Sciences*, 19(4), 394–407.

Wise, R. A., & Rompre, P. P. (1989). Brain Dopamine and Reward. *Annual Review of Psychology*, 40, 191–225.

Wood, W., & Neal, D. T. (2007). A New Look at Habits and the Habit-Goal Interface. *Psychological Review*, 114(4), 843–863.

World Economic Forum. (2020). *Ethics by Design: An Organizational Approach to Responsible Use of Technology*, www3.weforum.org/docs/WEF_Ethics_by_Design_2020.pdf

World Health Organization. (2019). *International Statistical Classification of Diseases and Related Health Problems, 11th Revision (ICD-11)*, Geneva.

Wu, J.-H., Robinson, S., Tsemg, J.-S., Hsu, Y.-P., Hsieh, M.-C., & Chen, Y.-C. (2023). Digital and Physical Factors Influencing an Individual's Preventive Behavior During the COVID-19 Pandemic in Taiwan: A Perspective Based on the S–O–R Model. *Computers in Human Behavior*, 139, 107525.

Xu, L., Feng, J., & Yu, L. (2022). Avalanche Criticality in Individuals, Fluid Intelligence, and Working Memory. *Human Brain Mapping*, 43(8), 2534–2553.

Xu, S., Yang, H., Menon, V., et al. (2020). Behavioral State Coding by Molecularly Defined Paraventricular Hypothalamic Cell Type Ensembles. *Science*, 370(6514). doi:10.1126/SCIENCE.ABB2494

Zeelenberg, M., Nelissen, R. M. A., Breugelmans, S. M., & Pieters, R. (2008). On Emotion Specificity in Decision Making: Why Feeling Is for Doing. *Judgment and Decision Making*, 3(1), 18–27.

Zink, C. F., Tong, Y., Chen, Q., Bassett, D. S., Stein, J. L., & Meyer-Lindenberg, A. (2008). Know Your Place: Neural Processing of Social Hierarchy in Humans. *Neuron*, 58(2), 273–283.

Zitek, E. M., & Tiedens, L. Z. (2012). The Fluency of Social Hierarchy: The Ease with which Hierarchical Relationships Are Seen, Remembered, Learned, and Liked. *Journal of Personality and Social Psychology*, 102(1), 98–115.

Index